P9-DNF-436

The Calculus of Friendship

∫ ∫ ∫

To Jill with my
best math wishes.
Don Joffray

The Calculus of Friendship

What a Teacher and a Student Learned about Life While Corresponding about Math

Steven Strogatz

PRINCETON UNIVERSITY PRESS ∫ PRINCETON AND OXFORD

Published by Princeton University Press, 41 William Street, Princeton, New Jersey 08540

In the United Kingdom: Princeton University Press, 6 Oxford Street, Woodstock, Oxfordshire OX20 1TW

Library of Congress Cataloging-in-Publication Data
Strogatz, Steven H. (Steven Henry)
The calculus of friendship : what a teacher and a student learned about life while corresponding about math / Steven Strogatz.
p. cm.
Includes bibliographical references and index.
ISBN 978-0-691-13493-2 (hardback : alk. paper) 1. Strogatz, Steven H. (Steven Henry)—Correspondence. 2. Joffray, Don, 1929– Correspondence. 3. Mathematicians—Correspondence. 4. Calculus. 5. Differential equations. 6. Chaotic behavior in systems. I. Joffray, Don, 1929– II. Title.
QA29.S698A4 2009
510.92′2—dc22 2008049339

British Library Cataloging-in-Publication Data is available

This book has been composed in

Printed on acid-free paper. ∞

press.princeton.edu

Printed in the United States of America

1 3 5 7 9 10 8 6 4 2

Dedicated to Mr. Joffray,
and to teachers like him everywhere

∫ ∫ ∫

Contents

Prologue

For the past thirty years I've been corresponding with my high school calculus teacher, Mr. Don Joffray. During that time, he went from the prime of his career to retirement, competed in whitewater kayak at the international level, and lost a son. I matured from teenage math geek to Ivy League professor, suffered the sudden death of a parent, and blundered into a marriage destined to fail.

What's remarkable is not that any of this took place—such ups and downs are to be expected in three decades of life—but rather that so little of it is discussed in the letters. Instead, our correspondence, and our friendship itself, is based almost entirely on a shared love of calculus.

It never occurred to me how peculiar this is until Carole (I'm happily remarried now) teased me about it. "You've been writing to him for thirty years? You must know everything about each other." Not really, I said. We just write about math problems. "That is such a guy thing," she said, shaking her head.

Her question got me thinking. What did I really know about my teacher? Why had so much gone undiscussed between us? On the other hand, we both enjoyed our correspondence the way it was, so was there any problem here?

Questions like these have kept nagging at me. I'm not sure how to go about answering them or if I should even try. All the while, I find myself looking for clues in a green

Pendaflex folder in my office, stuffed four inches thick with letters about math problems.

∫ ∫ ∫

I was 15 when I took calculus from Mr. Joffray. One thing about him was unlike any other teacher I'd ever had: he worshipped some of his former students. He'd tell stories about them, legends that made them sound like Olympian figures, gods of mathematics. In my own case, he was more a fan than a teacher, always marveling at what problems I could invent and solve. It felt slightly strange to be so admired by my own teacher. But I can't say I minded it.

After I graduated, something in me wanted to stay in touch with him. My first letters were about math problems that I thought he'd enjoy, gems I'd picked up in my college courses. The letters were infrequent, about one a year. I suppose he must have written back to me, but none of his responses have survived. It never occurred to me to save them.

It was only a decade later, when I was just starting my career as a professor, that our correspondence began to flourish. The pattern was always the same: Mr. Joffray would write to ask for help with a problem that had stumped him, typically a question raised by one of his seniors in the most advanced math class at the school. When one of these letters arrived in the mail, I stopped whatever I was doing to see if I could help. For one thing, they posed fascinating little questions, beautiful excursions off the beaten track of calculus. But maybe more importantly, they gave me a chance to explain math to someone who loved learning it, the best student any teacher could have, someone with perfect preparation and an evident sense of delight and gratitude.

With his retirement a few years ago and no more students to stimulate him, our correspondence began to wane. Not in frequency—in fact, he wrote to me more than ever—but in intensity and reciprocity. It got to the point where I simply couldn't keep up with him. Yes, he reassured me, he understood all that, and urged me not to worry; he knew how busy I must be in my career and with all the new obligations that come with raising a family. But it still felt like we were drifting apart. Ironically, I was now the same age that he was when he taught me in high school.

In January 2004, yet another letter arrived. But this time I felt anxious when I saw the envelope. The uncharacteristically tremulous handwriting reminded me of my dad's after his Parkinson's had set in.

Sat. January 17, 2004

Dear Steve,

Eek! I had a mild stroke Thurs. noon and lost all sensation in my right (writing) hand. Several hours later I managed to open and close my fingers and get some strength back into my grip, but, alas, no dexterity! X@#! A one-handed piano player isn't in demand, so I'll miss my gig with our jazz quartet tomorrow. . . .

This glimpse of mortality awakened me to how much I'd been overlooking all these years. I felt compelled to visit Mr. Joffray at his home, to come to know the man behind the math.

∫ ∫ ∫

Calculus is the mathematical study of change. Its essence is best captured by its original name, "fluxions," coined by its inventor, Isaac Newton. The name calls to mind systems that are ever in motion, always unfolding.

Like calculus itself, this book is an exploration of change. It's about the transformation that takes place in a student's heart, as he and his teacher reverse roles, as they age, as they are buffeted by life itself. Through all these changes, they are bound together by a love of calculus. For them it is more than a science. It is a game they love playing together—so often the basis of friendship between men—a constant while all around them is in flux.

The Calculus of Friendship

∫ ∫ ∫

Continuity (1974-75)

Calculus thrives on continuity. At its core is the assumption that things change smoothly, that everything is only infinitesimally different from what it was a moment before. Like a movie, calculus reimagines reality as a series of snapshots, and then recombines them, instant by instant, frame by frame, the succession of imperceptible changes creating an illusion of seamless flow.

This way of understanding change has proven to be powerful beyond words—perhaps the greatest idea that humanity has ever had. Calculus enables us to travel to the moon, communicate at the speed of light, build bridges across miles of river, halt the spread of epidemics. Without calculus, modern life would be impossible.

Yet in another way, calculus is fundamentally naive, almost childish in its optimism. Experience teaches us that change can be sudden, discontinuous, and wrenching. Calculus draws its power by refusing to see that. It insists on a world without accidents, where one thing leads logically to another. Give me the initial conditions and the law of motion, and with calculus I can predict the future—or better yet, reconstruct the past.

I wish I could do that now. Unfortunately, my correspondence with Mr. Joffray is riddled with discontinuities. Letters were lost or discarded. Those that remain are fragmentary and emotionally muted, and sometimes prone to half-truths, silver linings, and deliberate omissions.

$\int \int \int$

It was my sophomore year, spring term 1974. I was taking precalculus with a different teacher, Mr. Johnson, an MIT graduate, a tall, stern man, about 35 or 40, very fair but not given to smiling.

Some of my friends were in Mr. Joffray's section of the same class. I'd never talked to him and did not know much about him. There were rumors he'd been the national champion in whitewater kayaking. He was physically impressive—anyone could see that—big chest, muscular arms and legs, close-cropped hair. He looked like a stronger version of Lee Marvin, whom I'd seen in lots of war movies.

When we were learning about the rigorous definition of continuity—a very fundamental, difficult concept in calculus—Mr. Johnson told us something I'd never heard a teacher say before. It was ominous. He said he was going to present some ideas we wouldn't understand, but we had to go through them anyway. He was referring to the $\varepsilon - \delta$ definition of continuity:

> A function f is continuous at a point x if, for every $\varepsilon > 0$, there exists a $\delta > 0$ such that if $|x - y| < \delta$, then $|f(x) - f(y)| < \varepsilon$.

He said we'd need to see this four or five times in our education and that we'd understand it a little better each time, but there has to be a first time, so let's start.

It *was* difficult. Kids in our class were having a lot of trouble following the logic of these $\varepsilon - \delta$ arguments.

And then word filtered back to us that Mr. Joffray was doing things very differently in his class. He wasn't even trying to explain ε and δ. He'd defined a continuous function as one whose graph you could draw without lifting your pencil from the paper.

That told me a lot. Of course that was the intuition of what "continuity" must mean. But to leave it at that struck me, with my sophomore mentality, as taking the easy way out. It was soft. It was avoiding the issue. And so I began with a suspicion about Mr. Joffray, that he wasn't really hard core. I was glad I was in Mr. Johnson's class.

∫ ∫ ∫

The following year, Mr. Joffray was my teacher. Now I was able to take the measure of the man up close. Again I was impressed by his sheer physicality. His hands were the biggest I'd ever shaken—my hand was engulfed by his. And when he'd write on the blackboard, the chalk would pulverize with each stroke. Shards and splinters would fly off. Smithereens and dust all over him by the end of class.

He seemed to be very outdoorsy (something that had never appealed to me—I played tennis and basketball but never liked the woods—too many bugs—or canoeing or backpacking or any of that). A yearbook photo captured Mr. Joffray in his preferred habitat: high in a tree, inspecting a birdhouse he'd built. He was also faculty advisor to a group called the Darwin Club. I have no idea what they did, but it was outdoorsy nature stuff.

Anyway, what was his class like? Fun and pleasant. Low key. He was a happy man, friendly, always enthusiastic, though about strange things. He'd stride in and start talking about a goat that is tethered to a tree by a long rope. The stubborn animal pulls the rope taut and tries to walk away from the tree but ends up wrapping itself around the tree in

a tighter and tighter spiral. And then he'd ask us to find an equation for the goat's spiraling path.

I didn't know what to make of it. This wasn't suave, serious Mr. Johnson with his impressive MIT background. I felt like I was being taught by some sort of person I couldn't recognize.

But he was certainly jovial, so it was not an issue.

The math itself was interesting and came easily. I could learn it all from the book. His class did not add much, except for the weird nature problems.

On those signature occasions when he'd interrupt the class to tell us about his best former students, invariably he'd be in the middle of a calculation and then lapse into a reverie, with a faraway look in his eye and a smile breaking out. Then, in a hushed tone, he'd tell us about the time that Jamie Williams wrote down a formula for the nth term of the Fibonacci sequence.

Actually, that achievement did deserve reverence. As you may remember, the Fibonacci sequence is 0, 1, 1, 2, 3, 5, 8, 13, 21, 34 . . . . The sequence starts with the numbers 0 and 1, and after that each number is the sum of the two before it. The problem is, if $F_0 = 0$ and $F_1 = 1$, find a formula for F_n, the nth term. You'd want such a formula if you were interested in F_{100} or F_{1000}; you wouldn't want to have to add up a hundred or a thousand intermediate terms to get the answer. So is there a short-cut formula that expresses F_n directly in terms of n? The answer is amazing:

$$F_n = \frac{(1+\sqrt{5})^n - (1-\sqrt{5})^n}{2^n\sqrt{5}}.$$

How did Jamie Williams ever come up with that?

∫ ∫ ∫

With the passage of time I see now that I was like the goat tethered to the tree—and Mr. Joffray was the tree. I pulled taut on the rope and tried to get away from him, but only ended up wrapping myself closer and closer to him, all these years.

How did that happen? It wasn't because he taught me so much in the usual sense. No, his approach was so humble and unconventional, it confused me. It made me feel superior to him. I'm embarrassed to admit that, but it's true.

Here's what he'd do.

He'd suggest a problem, very gently, not at all insistent, and then he'd step aside. Often Ben Fine and I would compete to see who could solve it. Or if we could both solve it, who could solve it *better*.

Ben was a brilliant kid, a year younger than me, owlish and small, with sophisticated interests. (I often felt cloddish next to him.) And his math style was that of a languorous genius. He'd ponder the question without writing anything—he was a *philosophe*. Then, with a lightning stroke, he'd write a few lines of equations, render a poor sketch or two, and bam! He'd solved it.

Whereas I was a grinder. Not nearly as clever as Ben (and in retrospect, I see that he had much more talent for math than I did). My style was brutal. I'd look for a method to crack the problem. If it was ugly or laborious, with hours of algebra ahead, I didn't mind because the right answer was guaranteed to emerge at the end of the honest toil. In fact, I loved that aspect of math. It had justice built into it. If you started right and worked hard and did everything correctly, it might be a slog but you were assured by logic to win in the end. The solution would be your reward.

It gave me great pleasure to see the algebraic smoke clear.

And there was another reward. Mr. Joffray was an incredible cheerleader. He'd sometimes watch me and Ben, the tortoise and the hare, with a look of such admiration, almost awe, and happiness too.

At the end of my junior year, the school held its annual award ceremony. My name was called when they announced the Rensselaer prize for the top junior in math and science, and if I remember right, Mr. Joffray made a speech about me. He portrayed me as a mountain climber, ascending the mathematical peaks and then returning with tales of what I'd seen.

He made me sound generous, and heroic.

Pursuit (1976)

Having finished the math courses offered by the school, I spent my senior year teaching myself, detached from Mr. Joffray. I'd sit alone for an hour every day in an empty classroom, reading a textbook on multivariable calculus or trying to rederive Huygens's results about cycloidal pendulums.

At other times I did a kind of research, often about chase problems. These problems of pursuit, as mathematicians call them, completely captivated me.

The first one I ever heard of came from Mr. Joffray. It went like this: suppose a postman is trying to escape from a dog chasing him. The postman starts at the origin and runs away in a straight line at a constant speed v. Meanwhile, the dog starts at a point somewhere off that line and runs with constant speed w, instantaneously swerving in such a way that it is always heading straight toward the postman's current location. Find the equation for the curve traced out by the dog.

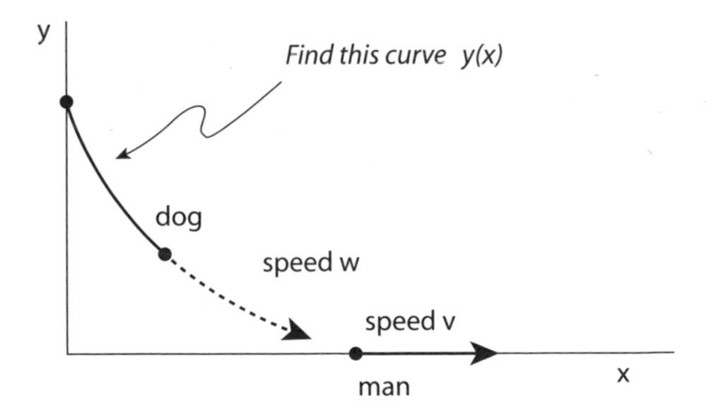

Another one was quintessential Joffray: a man in a kayak, paddling across a river, is trying to reach a certain point on the opposite shore. The kayaker, being a determined and not-too-clever sort, always aims directly toward his destination even as he's being carried downriver by the current.

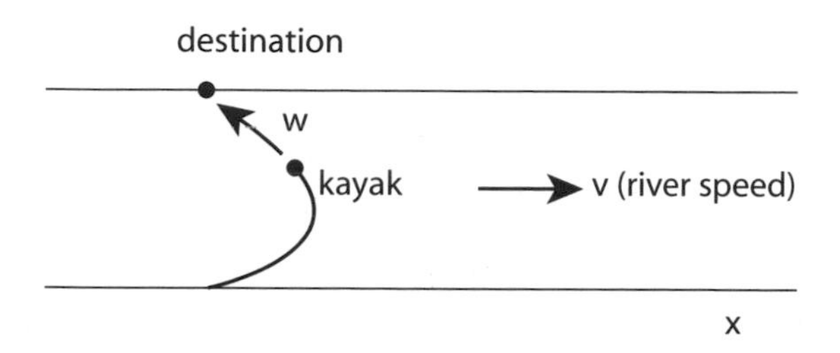

Suppose that the river flows with constant speed v and the kayaker paddles with constant speed w relative to the river. Find the path traced out by the kayak.

These two chase problems—the dog and postman, and the determined kayaker—are both exercises in differential equations. Such equations are what calculus is all about: flux and change. A differential equation describes how a system changes its behavior in response to the ever-changing forces on it. All the pushes and pulls nudge the system to be in some new condition, or some new place, where the forces are different again. For instance, in the dog-vs.-postman problem, the man keeps moving, so the dog has to keep revising its directional heading. It does that *instantly*.

That's the breathtaking idea behind all of calculus—you think about what's happening at an instant, in an infinitesimal unit of time—and you can actually deal with that ineffable idea, make it into a powerful predictive tool. Here we can write down a differential equation that captures the idea of "aiming," of the fact that the dog changes its course heading at every instant. And by *solving* that equation, we learn the whole path that the dog must follow. The whole trajectory is built of all the infinitesimal steps the dog takes en route to its prey.

This vision of the world—that everything can be viewed as the accumulation of infinitesimal changes—is the most revolutionary insight of calculus. Figuring out how to turn this idea into workable mathematics was the breakthrough that allowed calculus to be invented in the first place, back in the 1600s. Isaac Newton was trying to calculate how the planets move. He did this by thinking of the planets as being acted on by the ever-changing force of gravity. As they orbit the sun, they change their distance from it, which changes the gravitational tug they feel, which then steers them to a new place in the next instant, where the force is again slightly different, and so on. Solving for the motion of the planets becomes a problem in differential equations.

So by playing with chase problems, you can feel like you're in the company of Newton. And you *are*. I loved that about them.

$\int \int \int$

The chase problems that Mr. Joffray had tossed out were challenging but ultimately manageable. Their compliant character reinforced my feelings about mathematical justice. All you had to do was translate the word problem into the right equations; find the right substitutions; work calmly and logically and grind though the algebra without doing anything wrong; and sure enough, the right answer would pop out. It had to.

The first hint that something was wrong with this gauzy vision came when I concocted a question of my own. It seemed a lot like the other chase problems I'd solved, but for some reason it was turning out to be strangely obstinate. I spent months on it. It was frustrating, tantalizing, and delicious. With enough effort, I felt sure I could solve it, and all the months of frustration would make the conquest that much sweeter.

The question was, suppose a dog at the center of a circular pond sees a duck swimming around the circumference. The dog chases the duck by always swimming straight toward it. In other words, the dog's velocity vector always lies along the line connecting it to the duck. Meanwhile, the duck takes evasive action by swimming around the edge of the pond as fast as it can, always moving counterclockwise. Assuming both animals swim at the same constant speed, find an equation for the dog's path.

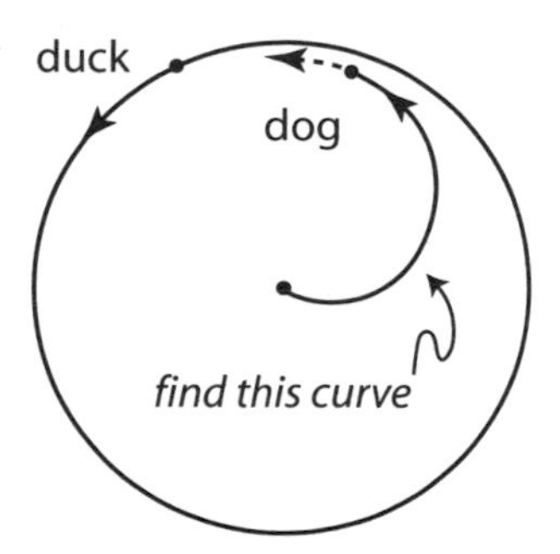

It's clearly some sort of spiral that asymptotically approaches the circle where the duck is. But what's the equation for such a spiral?

Unlike the spiral of the tethered goat winding itself tighter and tighter around a tree, this is an *expansive* spiral, growing away from its center but limited by a circular boundary it can never cross.

A fascinating path.

I couldn't calculate it. I tried every conceivable change of variables and even managed to reduce the problem to a beautiful differential equation, one that looked like it should be tractable. But I could never solve that one either.

What I didn't know yet was that some math problems are unsolvable. No explicit answer is possible; in this case, there's no formula for the dog's spiraling path. It's indescribable in terms of the elementary mathematical functions we're used to. You might say our language isn't up to the task.

In later years I'd realize that this is the rule, not the exception. Most differential equations are unsolvable in the same sense. Our library of formulas isn't rich enough to encompass them. Which makes those few problems we *can* solve—the ones they give you in high school—seem all the more precious.

Relativity (1977)

The theory of relativity is founded on empathy. Not empathy in the ordinary emotional sense; empathy in a rigorous scientific sense. The crucial idea is to imagine how things would appear to someone who's moving in a different way than you are.

At a time when it seemed absurd to claim that the earth moves around the sun, Galileo asked doubters to imagine being confined below deck in an enormous ship. There are no portholes in your cabin, no way to see the coastline passing by. If the sea is calm and the boat is gliding straight ahead at a steady speed, how could you tell you are moving? You couldn't. Any observation you'd make would be the same as if you were motionless. Pour wine into your glass, and it falls straight down, just as it would on dry land. That's because everything in your cabin—the furniture, the air, the wine—would be moving right along with the ship. By the same reasoning, said Galileo, the earth could be moving and we wouldn't sense it.

Three hundred years later, Einstein wondered what he would see if he were riding alongside a beam of light. Would the electromagnetic waves stand still, in violation of Maxwell's equations? Would a distant clock seem to stop ticking? From questions like these came the special theory of relativity, with its shattering insights about time and space, mass and energy. Later he asked how the laws of physics would appear to an observer inside a plummeting elevator (assuming he retained his composure long enough to make observations).

As a teenager I revered Einstein. For his genius, of course, but also because he radiated kindness. Silly as it sounds, he was a big part of why I wanted to attend Princeton. I wanted to be near him, to walk in his footsteps. Soon after arriving, I dragged a few other freshmen along with me to gaze at his house.

Lately, though, I've started to see him differently. He wasn't perfect. There was even something slightly sad about him. Though a playful man in many ways, he was aware of missing out on the deepest forms of human intimacy. "I am truly a 'lone traveler' and have never belonged to my country, my home, my friends, or even my immediate family, with my whole heart," he once wrote. For all his empathy in the abstract, Einstein remained curiously detached from the people closest to him.

∫ ∫ ∫

My correspondence with Mr. Joffray began with a letter dated March 26, 1977, written in the spring of my freshman year of college. I'm not sure what triggered it. After all, he and I had fallen out of touch during my senior year of high school, while I worked alone on chase problems. And despite how much I'd enjoyed his calculus class the year before that, it hadn't brought us especially close. He hadn't ever been my mentor or trusted adviser; another teacher, Mr. DiCurcio, was. And, other than calculus, I wasn't the least bit interested in the things that interested Mr. Joffray: nature, team sports, water sports, etc. Still, something made me write to him that first time. What was it?

As I reread the letter now, I see that I was hiding more than I was revealing when I wrote "Princeton is all I've ever dreamed of. (Except for my math instructors so far . . .)."

What I couldn't bring myself to admit was that my first math course in college had utterly deflated me and changed the way I viewed myself. It was a proof-oriented course on linear algebra, aimed at freshmen who had aspirations of being math majors. It was intended to be a taste of rigorous, abstract math—the sort of thing you'd have to be good at if you wanted to be a pure mathematician. The professor, a renowned topologist, was so shy that he slithered along the wall when he entered the lecture room on the first day, as if hoping to become invisible. He spent the rest of the semester looking down at his Wallabees and tugging at his red beard. The few times I dared to ask him a question, he seemed startled and gave a monosyllabic response. I read the book, did the homework, paid careful attention in lecture, and had no idea what was going on. It was terrifying. No matter what I did, I couldn't get it. The textbook was dry, hyperprecise, and devoid of illustrations. The homework was baffling. And the tests—just thinking about an upcoming test would send me running to the bathroom.

Understandably, I spared Mr. Joffray the details. Even the single crumb of small talk I offered in the letter was apparently too much; after apologizing for "stalling," I changed the subject back to math, specifically, chase problems—like Mr. Joffray, an old friend from a happier time.

The mathematical theme of the letter was that changing your frame of reference can be powerful. Sometimes, as in relativity theory, a frustrating problem becomes clearer when viewed in the right frame.

∫ ∫ ∫

The question discussed in the following letter is a chase problem about four dogs. Starting from the corners of a square of side a, each dog chases the one counterclockwise from it. If they all start at the same time and run at the same speed, how far has each dog run by the time they all collide at the center of the square?

It looks hard. All the dogs spiral in as each chases the one in front of it. The problem is to find the arc length of that spiraling path.

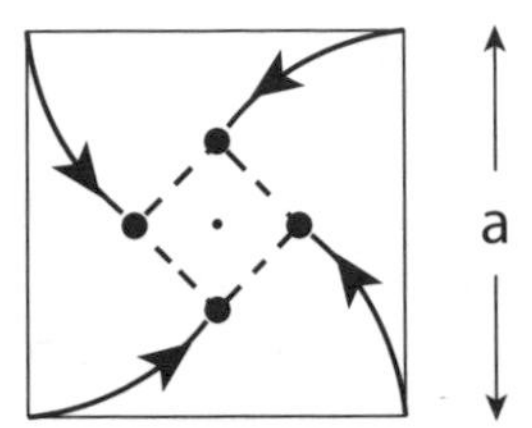

The conventional way to solve this problem uses calculus. Think about where the dogs are at any given time. By the symmetry of the problem, they will always be at the corners of a square somewhere inside the original square and centered at the same point. The little square will shrink and rotate as the dogs chase each other.

Look at one of the dogs at a corner of that shrinking square. Call its distance from the center r, where we're now thinking in terms of polar coordinates.

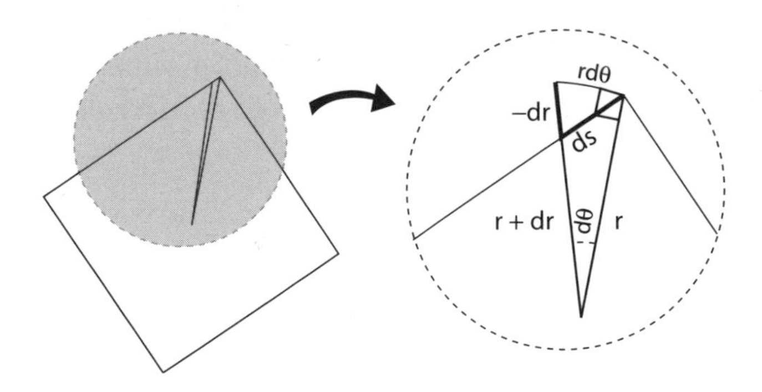

In the next instant dt, the dog moves a distance ds in the direction of its target (the dog it's chasing). That little element of arc length, ds, can be viewed as the hypotenuse of an infinitesimal right triangle with sides of length $rd\theta$ (here, we pretend $rd\theta$ is a straight line segment, not an arc of a circle) and $-dr$. Notice that we have to write $-dr$ because dr is negative; the dog's motion *decreases* its distance r from the center in the next instant, as the picture shows.

Looking at that infinitesimal triangle, we see it's actually an *isosceles* right triangle with $\frac{\pi}{4}$ as the base angles. So

$$\tan(\pi/4) = \frac{-dr}{rd\theta}.$$

But since $\tan(\pi/4) = 1$, the equation for the dog's path $r = r(\theta)$ satisfies

$$-\frac{dr}{rd\theta} = 1.$$

This differential equation is easy to solve. Just separate the r's from the θ's and integrate both sides: $\int \frac{dr}{r} = -\int d\theta$, which implies $\ln r = -\theta + C$, or equivalently,

$$r = Ke^{-\theta},$$

where $K = e^C$. This is the equation of a "logarithmic spiral," a famous and beautiful shape first studied by the Bernoulli brothers, two of the greatest mathematicians in the generation after Newton.

To evaluate the constant K, note that at $\theta = 0, r = \frac{a}{\sqrt{2}}$ (if we start the problem with the dogs in a diamond shape like this):

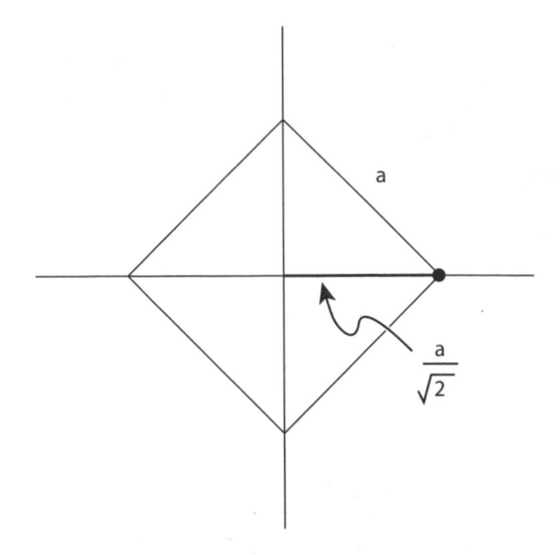

Thus $r(\theta=0)=\frac{a}{\sqrt{2}}$, but since $r=Ke^{-\theta}$, we get $K=\frac{a}{\sqrt{2}}$. Hence

$$r=\frac{ae^{-\theta}}{\sqrt{2}}$$

is the curve traced by one of the dogs.

To finish the problem, we need to figure out how far this dog has run when it collides with the others at the center of the square. That happens at $r=0$, which means $\theta\to\infty$ (an infinite amount of spiraling has taken place). The arc length from $\theta=0$ to $\theta=\infty$ is found as follows: from the isosceles right triangle, we know $ds=\sqrt{2}\,rd\theta$; this comes from the Pythagorean theorem, where we think of ds as the hypotenuse and $rd\theta$ as the base. Now substitute for r, using $r=\frac{ae^{-\theta}}{\sqrt{2}}$, to obtain $ds=ae^{-\theta}d\theta$. Therefore

$$s=\int ds=a\int_0^\infty e^{-\theta}d\theta=a(-e^{-\theta}|_0^\infty)=-a[0-1]=a.$$

That's a suspiciously simple answer: each dog has run a total distance a, the side of the square, when it crashes into the other three at the center of the square.

The following letter shows how to get this answer without using calculus.

Dear Mr. Joffray, March 26, 1977

I have a real little gem for you this time! Remember the following problem? (A chase problem of course!) Four dogs chase each other—each starts from the corner of a square and runs directly toward the one counterclockwise from it. How far has each travelled when they meet in the center?

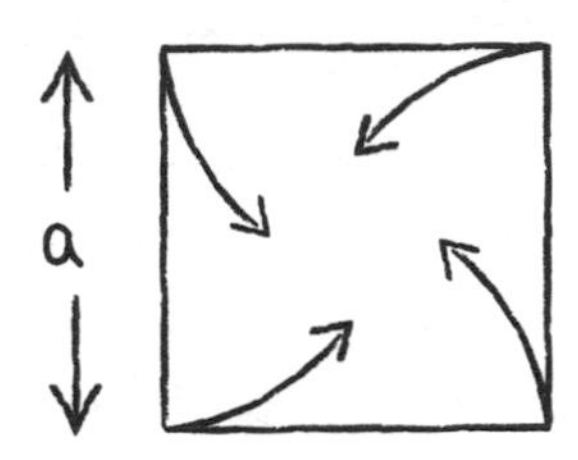

You may remember the remarkable solution: distance travelled $= a$! Then I generalized the problem to n dogs starting from a regular n-gon. This problem also has a nice, compact answer:

$$d = \frac{a}{2}\csc^2\frac{\pi}{n},$$

where a is the length of a side. All this is ancient history. But remember I once mentioned the possibility of an intuitive solution (although it evaded Ben Fine, so the possibility seemed slim indeed!). Well, you have guessed right if you think I'm going to show you an intuitive solution. But first . . .

I hope you're well and that Loomis is finishing off another successful year. As for me, my glee has not diminished—Princeton is all I've ever dreamed of. (Except for my math instructors so far. Two out of three have been mathematicians, but not teachers; they tried to drown me with rigor in my second-semester Advanced Calculus course, but fortunately I swam off to the familiar shores of a "Math 6" level course. Linear Algebra (first semester) was also too abstract, but I felt obligated to give it a chance. I have a feeling this sort of abstraction killed Jamie Williams's interest in math, and I wasn't going to let that happen to me. So all is peachy again.)

I've been playing a lot of tennis and even generating a little social life for myself.

OK, enough stalling.

First, let's start with the four dogs. Would you agree that, by symmetry, all the dogs are always in a square in relation to each other? This means that, at any instant, the dogs' velocities are at right angles. Now consider the segment connecting the positions of any two dogs.

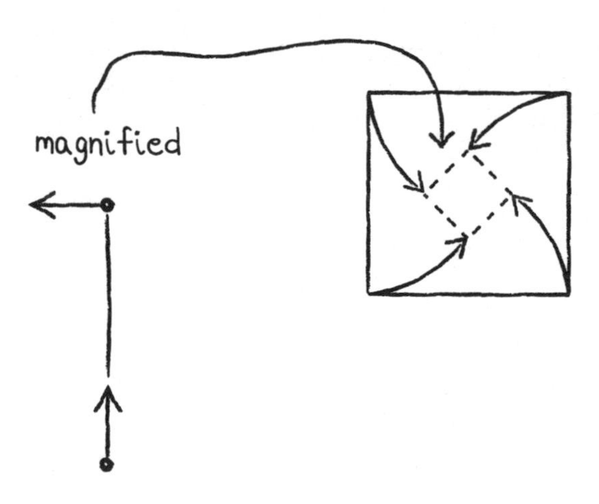

(This is the tricky part to phrase, coming up.)

At this instant, *along the line segment that connects them*, the chasee has no velocity component, whereas the chaser has *all* its velocity directed along that line. So if we just think about the connecting segment at all the instants, it is as though the chasee is never "running away" (in the sense that its velocity is normal to the segment). The chasee may as well stand still in its corner—that's how much good its sideways motions will do it. The chaser is resolute and homes in steadily. Do you see that the distance travelled by the chaser will be a?

If you don't, here's an alternative but equivalent way of looking at it. Put a movie camera on the chaser's head. It swivels in such a way as to always keep the chasee in the center of the picture. (Actually, that's wrong—it won't have to swivel at all, will it?)

Anyway, put this camera on his head, roll 'em, and let the chase begin. After the chase, roll the film and show the result on a screen. What will you see?

The chasee will appear in the center of the picture and won't seem to be running away. You're just moving closer. Could you distinguish this movie from the following one (assuming no background!)? The chasee stays still in its corner, the chaser runs a and catches the chasee? You couldn't distinguish, so in both cases, a is the answer. (Of course, all this camera talk is a conceptual scheme to eliminate the *normal* component of the chasee's velocity.)

As an exercise, I'll leave you the n-gon problem. I suggest you keep the camera approach in mind. You should get $d = \frac{a}{2}\csc^2\frac{\pi}{n}$. (Sorry, Mr. Joffray—I couldn't resist leaving *you* an exercise!)

In friendship,
Steve

P.S. Martin Gardner, of *Sci. Amer.* fame is responsible for the approach mentioned above.

P.P.S. Give the problem but not the solution to Ed Rak. Let's see what he comes up with.

Irrationality

(1978-79)

After my demoralizing encounter with linear algebra, I thought about switching from math to physics. Then in my sophomore year I had an inspirational teacher, Elias Stein, for complex analysis, and so I went back to majoring in math again.

At some point during that year, my brother Ian had a talk with me. It's hazy in my memory, but I picture us driving home for Thanksgiving. He was asking me about my future plans, what was I going to major in, etc., and then started trying to persuade me to take all the pre-med courses. Not this again—everyone had always told me I should be a doctor. "You like math and science," they'd say. "And there are so many things you can do with it—some are even mathematical, like radiology." ("And you have such nice hands," my mother would say.)

But Ian took a different tack. He was a sharp lawyer and a great judge of people. Plus, he truly did think he was giving me good advice. He argued that I should take biology and chemistry and organic chemistry because I might actually like them. It wouldn't commit me to becoming a doctor. It would just be so much easier to take them now rather then jam them in later. And even if I didn't end up going to medical school, I'd still benefit from the broad science background they'd provide.

So after a lot of soul searching, I relented and gave pre-med a try.

Of course this was completely irrational. I had *no* interest in being a doctor. I wanted to be a math professor, and always had.

All of that must have been very much on my mind when I wrote to Mr. Joffray in February of 1979. But I barely mentioned it to him. After all, he was not my confidant or coach. He was someone who enjoyed my math excursions, and I guess I thought those were the rules: I'd share math problems with him.

∫ ∫ ∫

The letter was about the irrationality of $\sqrt{2}$. Mathematicians care about this number because it's fundamental to geometry. It tells you how long the diagonal of a square is compared to its side. For instance, a $1' \times 1'$ square has a diagonal that's $\sqrt{2}$ feet long.

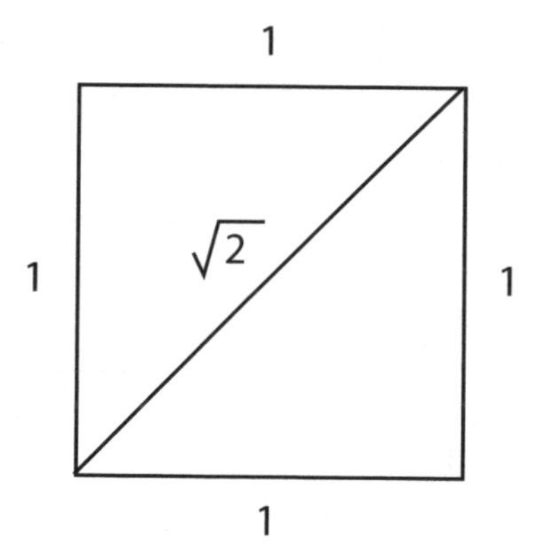

That fact comes straight from the Pythagorean theorem, which says that a right triangle with sides of length a and b has a hypotenuse of length c, where $a^2 + b^2 = c^2$. Here, we make the connection by regarding the diagonal of the square as the hypotenuse of a right triangle with sides $a=1$ and $b=1$. Then we get $c^2=1^2+1^2=2$, and hence $c=\sqrt{2}$ is the diagonal's length.

In ancient times, the Greeks, and probably the Babylonians and maybe the Indians and the Chinese, knew that the diagonal of a square was about 40% longer than its side, but what exactly was this basic and magical number $\sqrt{2}$?

The Pythagoreans originally suspected that $\sqrt{2}$ must be the ratio of two whole numbers, and they desperately wanted to know what those two numbers might be. This was no minor curiosity to them; it was part of an all-encompassing religion and research program founded on the mystical idea that "all is number," that the laws of the universe could be expressed in mathematics, and more specifically, in terms of small whole numbers. Pythagoras himself had made the discovery that musical harmony was based on numbers: when he plucked two lyre strings, both strung at the same tension but one twice as long as the other, the longer string sounded a note exactly one octave lower than the shorter string. In fact, any string ratio involving small numbers, such as 2:3 or 4:3 or 5:2, produced beautiful harmony.

With this in mind, it was natural to expect that $\sqrt{2}$ must also be the ratio of some nice whole numbers. It's almost $\frac{7}{5}$, but not exactly, because $5^2+5^2=25+25=50$, whereas $7^2=49$. Close, but not quite. Maybe two bigger numbers would work?

That's the question. Can we find two whole numbers, m and n, such that $\sqrt{2}=m/n$, or equivalently, $n^2+n^2=m^2$?

The shocking answer is no. The square root of 2 is *irrational*, meaning it can't be written as the ratio of two whole numbers. The philosophical implications were profound and disconcerting. If whole numbers were inadequate to characterize something as simple as a square's diagonal, what hope was there for the rest of Pythagorean philosophy?

The legend goes that when one of Pythagoras's disciples discovered the irrationality of $\sqrt{2}$, the cult members were so enraged that they took him out to sea and hurled him overboard.

∫ ∫ ∫

There's a standard proof that $\sqrt{2}$ is irrational. You've probably heard people describe certain pieces of mathematics as elegant. Well, this proof is, in my opinion, not elegant, whereas the unorthodox one in the letter to Mr. Joffray is. See if you agree.

The usual proof goes like this. Suppose $\sqrt{2} = m/n$, where m and n are whole numbers with no common divisors. In other words, before doing anything else, we divide out any factors common to m and n so that the fraction is in "lowest terms." We do that for cleanliness. It ensures that m and n will be as small as possible, and it avoids the confusion that could ensue if we allowed a number to be represented by several different fractions, such as $1/2 = 2/4 = 3/6$. By requiring lowest terms, we make sure that fractions are represented in a unique way.

Next we'll derive a contradiction, which will imply that it must be *impossible* to write $\sqrt{2} = m/n$ for any m and n.

Okay, here goes. If

$$\sqrt{2} = \frac{m}{n},$$

then

$m = \sqrt{2}n$	
$\Rightarrow m^2 = 2n^2$	(we squared both sides)
$\Rightarrow m^2$ is even	(because it's equal to 2 times something, where "something" is the whole number n^2)
$\Rightarrow m$ is even	because the square of an even number is always even and the square of an odd number is always odd; think about this!

So now we know m must be even, which means we can write it as 2 times something, say $m = 2p$, where p is a whole number.

Here comes the contradiction. Since $m = 2p$ and $m^2 = 2n^2$ (from the reasoning above), we see that $(2p)^2 = 2n^2$, which implies $4p^2 = 2n^2$. Cancel a factor of 2 to get $2p^2 = n^2$. And there's the rub. Because $2p^2 = n^2$, we see that n^2 is also 2 times something, which means n^2 is even and hence n is even.

Thus, we've concluded that both m and n are even. That contradicts our earlier requirement that m and n have no common divisors (we've just shown they're both even and hence both divisible by 2). This implies that our original assumption was wrong, and hence $\sqrt{2}$ must, in fact, be irrational.

The logic of this proof is airtight, but there's something annoying about it. Aside from its meandering character, the argument is unthematic. It makes the irrationality of $\sqrt{2}$ seem like a fact about number theory, not geometry. What happened to all the shapes we started with—the squares, diagonals, and triangles?

In contrast, the proof in my letter to Mr. Joffray is purely geometric. It was shown to me by one of my teachers at Princeton, Benedict Gross.

Dear Mr. Joffray, February 20, 1979

It's time for our annual math problem! I hope this note finds you and your family well and happy. Everything's fine from this end—my direction is now away from teaching and toward medicine, and the pre-med courses are stimulating and competitive. But I still love mathematics best (I'm majoring in it) and couldn't resist showing you this sly proof of an elementary fact in geometry: the irrationality of $\sqrt{2}$.

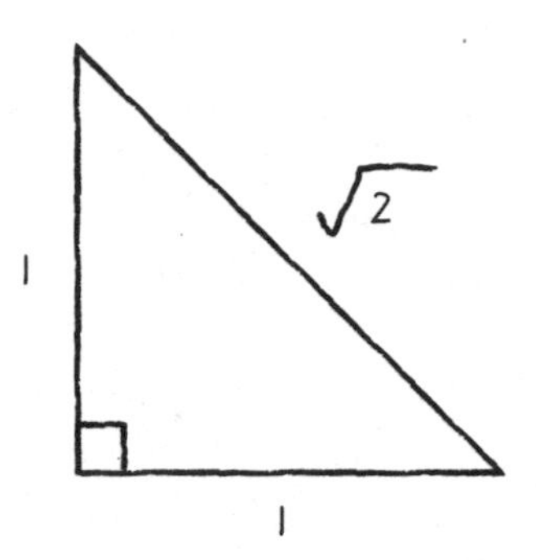

Suppose $\sqrt{2} = m/n$, where m and n are integers. Think of n in this way: it is the number of times we repetitively add some unit length until we achieve the length of a side.

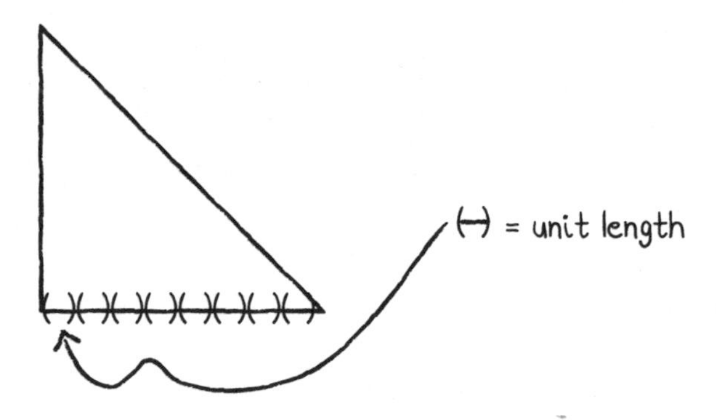

Maybe I'm being wordy: the sides are n times the unit length, and the hypotenuse is m times, OK? Now make the following construction: mark off a length DB equal to the length of AB. Now drop the perpendicular DE (to AC) and draw in the line EB.

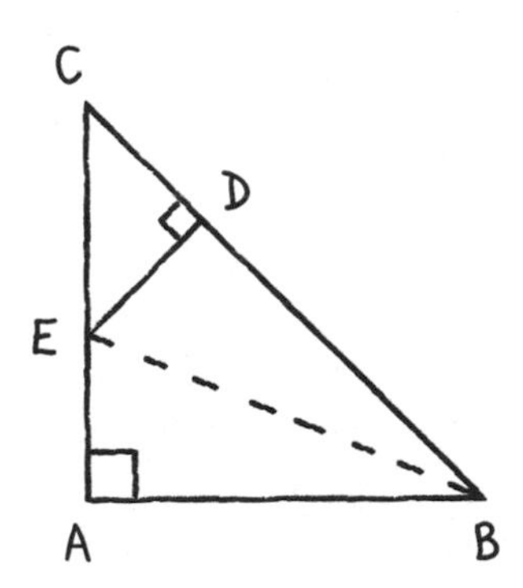

Claim: $\triangle ABE \cong \triangle DBE$. This follows because the triangles share $\overline{EB}$, they're both right triangles, and $\overline{DB} \cong \overline{AB}$ by construction.

Hence $\overline{EA} \cong \overline{ED}$ by corresponding parts; then $\triangle CDE$ is soon to be an isosceles right triangle—it has an angle $\angle ECD = 45^\circ$ and is a right triangle by construction. Thus $\overline{CD} \cong \overline{ED} \cong \overline{EA}$, as shown by the tick marks.

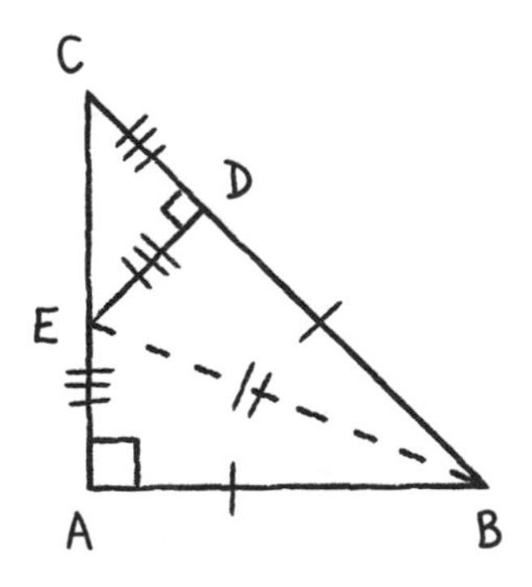

So far, nothing.

Now for the proof (I'm hoping you haven't seen this before!).

We assumed that the lengths of $\overline{AB}$ and $\overline{BC}$ could each be expressed as a multiple of some unit length, call it ℓ. Then consider the new, smaller isosceles right triangle CDE.

Look first at $\overline{CD}$. Its length = length of $\overline{CB}$ minus length of $\overline{DB}$; but $\overline{DB} \cong \overline{AB}$, so length of $\overline{CD}$ = (length of original hypotenuse) – (length of original side), both of which were integral multiples of the unit length ℓ. Hence their difference is also an integral multiple of ℓ (closure of the integers!!), and therefore we now have a smaller isosceles right triangle whose side $\overline{CD}$ is an integral multiple of ℓ. Wait—we must show the same fact is true of the little hypotenuse $\overline{CE}$. Its length is that of $(\overline{AC}) - (\overline{AE}) = (\overline{AC}) - (\overline{CD})$, both integral multiples of ℓ. ($\overline{AC}$ = original side; $\overline{CD}$ is shown above to be a multiple.)

So our construction yields a smaller isosceles right triangle whose sides and hypotenuse are both multiples of the unit length. (It's clear that the scale factor for length shrinkage is > 2; i.e., 2 times length $\overline{CE} <$ length $\overline{BC}$.)

Hence (do you see what's coming?) we can construct a sequence of isosceles right triangles of ever-diminishing size; in fact, we can construct a triangle of very small size *but whose sides and hypotenuse are still integral multiples of the unit length* ℓ (which follows from earlier arguments).

But this is crazy! If we choose a triangle with a hypotenuse of length $<\ell$, what are we to do? No integral multiple of ℓ (other than 0) is less than ℓ in length—and so we have a contradiction.

Thus no such ℓ could exist, and therefore the hypotenuse and the side are incommensurable (some nice old-fashioned language there!). This proof is due to my professor in a seminar on Fermat's last theorem (haven't solved it yet, by the way).

I prefer the proof to the usual one because it's more in the geometric spirit of the question. If you get the chance, tell me what you think, and also what's new.

So long,
Steve

P.S. Exercise for the reader: Use the method of this proof to show that the Golden Mean is irrational (i.e., the sides of the Golden Rectangle are incommensurable). Au revoir!

A turning point came soon after I'd mailed this letter.

Having been convinced by my brother to switch to pre-med, I was now shouldering a course load consisting of abstract algebra and differential geometry, plus freshman biology, freshman chemistry, and organic chemistry, all at the same time. Three labs a week. That's tough for anyone, especially someone so inept in real-world things. I was always the last to leave the orgo lab. The teaching assistant hated me. "What's taking so long," she'd ask. "It's just like boiling water, or cooking," she'd say. But I'd never done those things either.

Another drag was preparing for the MCAT, the entrance test required for medical school. Because I had so much to catch up on, I'd signed up for a Kaplan review course. The nearest one was in New Brunswick—an hour away by car—and it met on Sunday mornings. Not a time a college student wants to be awake, and least of all, preparing for the MCAT.

Still, I resigned myself to the pre-med conveyor belt.

When I came home for spring vacation, my mother took one look at my face and said, “There’s something wrong. Something’s really bothering you. How’s school?”

“I like it,” I said. “It’s fine, I’m learning good stuff.”

“No, you don’t look happy. What’s the matter?”

I didn’t really know. “Maybe I’m tired,” I said. “I’m working a lot.”

“No, it’s something else. What are you going to take next year? You’ll be a senior.”

Well, that *was* bothering me. “Being a pre-med so late, I’m going to have to take biochemistry, vertebrate physiology, and a bunch of other courses that the med schools want. Plus I have a senior thesis and two more courses in the math department. Which means that my schedule is going to be so full that I’m not going to be able to take quantum mechanics.”

“Why does that matter to you?” she asked.

“Because that’s something I’ve wanted to learn about forever!” I blurted out. “In first grade, when the teacher would march us down to the school library and make us choose a book, I’d always choose the same one: *The How and Why Wonder Book of Atomic Energy*. And it’s been like that ever since. Bohr and Heisenberg, Schrödinger and Einstein—those guys are my heroes. I’ve worked my whole life to get to this point, and now I’m finally ready to learn what the Heisenberg uncertainty principle really says—not just the words, but the math. But I’m never going to get the chance to do it because I can’t fit it in, and then I’ll be in medical school, cutting up cadavers, and it’ll be too late.”

She looked at me, caught my eyes.

“What if you could just say, right now, ‘I love math and physics. I want to take quantum mechanics. I’m not going to be a doctor. I'm going to do whatever it takes to become the best math professor I can be.' "

And I started to cry. It was as though a tremendous weight had been lifted. Then we were both laughing and crying. I never did take the MCAT, and I've never looked back on that decision. Some people never find their passion. But by denying mine, I found it and became sure of it.

Shifts (1980-89)

Once I committed to pursuing math, life took me on a straight trajectory for the next nine years. These were the years of training. During that time Mr. Joffray and I seldom wrote to each other. There are just three items in the green Pendaflex folder from that period. But a few shifts were already starting to take place. In a letter dated December 16, 1980, I thanked Mr. Joffray for the first time:

> So till we meet again, be well and continue your great work—did I ever thank you for your encouragement? Well, I do so now—it was a big help.

The letter also maps out a linear path for my life:

> I'm still planning to get a Ph.D (maybe Theoretical Physics, maybe Applied Mathematics) and—eventually—to get married (no candidates yet on the horizon), become a professor, learn for the rest of my life, and live more or less happily ever after.

What would Mr. Joffray have said about this straightforward view of life? Would he have corrected me and told me life is not linear, that it's more like the unpredictable currents of a whitewater kayak course? There's no way to know because his reactions to this letter, and to all my letters up until 1989, have been lost, which is a euphemism for what really happened. I threw them away.

In 1981 I wrote to him again. That letter is about shifts of a different kind: the mathematical shift operations that allow you to solve for the Fibonacci numbers in closed form, a rational pathway to Jamie Williams's legendary achievement.

Dear Mr. Joffray, 9/21/81

I was in the English Shop today in West Hartford, and I bumped into an old friend from Loomis. We had a quick chat, and I was reminded how much I miss you and the rest of the gang. I'll give you a quick personal update, and then share a math problem with you. I recall that you once expressed amazement at Jamie Williams's discovery of a closed expression for the nth term of the Fibonacci sequence. In this letter I'll show you a method for cranking out closed expressions for all types of sequences.

Actually I wonder how Jamie did it—probably just "saw" the answer and then used induction. Amazing!

Well, as for me, I graduated from Princeton in 1980 with an A.B. in Math. My senior thesis dealt with a problem in biochemistry that required some tricky math (topology and differential geometry). It had plagued the biochemists, but I managed to solve it (maybe!) and in any case published a paper in the *Proceedings of the National Academy of Sciences*.

I was awarded a Marshall Scholarship to study applied math at Trinity College, Cambridge University (Newton's old college), for two years. I'm going back for my second year on Wednesday—it's a terrific experience, culturally and academically. I even managed to play basketball for the Cambridge University team (my first time since Loomis JV!). After Cambridge, I hope to study applied math for a Ph.D. at either Harvard or Berkeley. And ultimately, with luck, I'll find a job as a college professor somewhere.

Hope all is well with you and your family. I enjoyed the Bulletin's write-up about you, even though it arrived about 5 months late (slow postage to England!)

Now it's time for some mathematics. I learned the method which I'll soon describe while teaching some gifted high school students at a summer math program at Hampshire College two summers ago.

Our strategy for studying sequences defined by recursion (i.e., if you know the first n terms, you know the $(n+1)^{\text{st}}$ term because of some defining rule) is to introduce an operator called the shift operator. When rewritten in these terms, our Fibonacci problem is simplified, as I'll show now.

The pleasant thing about this operator is that it gives us only a single subscript (i.e., have a_n but lots of operators around, rather than a_{n+2}, a_{n+26}, etc.). You'll see what I mean.

1. **The shift-left operator:**

Suppose we have a sequence $x_1, x_2, x_3, \ldots = \langle x_n \rangle$. Then the shift-left operator L is given by the rule

$$L(x_n) = x_{n+1},$$

where the notation is that x_n is an element of the sequence $\langle x_n \rangle$. So note that if L acts on the whole sequence

$$x_1, x_2, x_3, x_4, \ldots,$$

it yields the sequence $\langle L(x_n) \rangle = Lx_1, Lx_2, Lx_3, \ldots$

$$= x_2, x_3, x_4, x_5, \ldots,$$

which is just the original sequence shifted one notch backward (i.e., yanked back one space to the left). Hence the name "shift operator."

2. So the Fibonacci rule $a_{n+2} = a_{n+1} + a_n$ with $a_1 = a_2 = 1$ may be written in terms of our shift operator L:

$$a_{n+1} = L(a_n); \qquad a_{n+2} = L(a_{n+1}) = L(L(a_n)) = L^2(a_n).$$

So $a_{n+2} - a_{n+1} - a_n = 0 \Leftrightarrow L^2(a_n) - L(a_n) - 1 \cdot a_n = 0$. Symbolically, it's slick to rewrite this by "factoring out" a_n; i.e.,

$$(L^2 - L - 1)a_n = 0.$$

Just think of $L^2 - L - 1$ as a machine, or a black box, that takes a number a_n as its victim and generates a new number. If a_n happens to be a Fibonacci number, zap! It is annihilated, and zero is the result.

Note the analogy to differential equations where the derivative operator D sends the function y to its derivative y' ($Dy = y'$). I always thought it was fun and amazing to play with algebraic formulas involving D, as if it were just a variable, not an operator; e.g., it amazed *me*, at least, that

$$(D^2 + 1)y = 0 \quad \text{is the same as}$$
$$(D + i)(D - i)y = 0.$$

Even more incredibly, you can now pretend it's just algebra and solve $(D + i)y = 0$ and $(D - i)y = 0$ separately! Then the general solution is a "linear combination" of the two separate solutions Ae^{-ix} and Be^{ix}. Best of all, the answer comes out right!

3. **The key point** (for theoretical people).

All the tricks above are possible and legal because D is a *linear* operator; i.e., $D(y_1 + y_2) = Dy_1 + Dy_2$, and $D(ay) = aDy$.

The great thing for us now is that our shift operator L (which sends sequences to sequences rather than functions to functions) *is also linear.*

So all the machinery of linear analysis (which I won't bore you with) permits us to solve

$$(L^2 - L - 1)a_n = 0$$

by solving the factored versions of this equation separately.

Let ϕ and θ be the roots of the algebraic equation $m^2 - m - 1 = 0$. Then $m = (1 \pm \sqrt{5})/2$; let

$$\boxed{\phi = \frac{1+\sqrt{5}}{2}, \qquad \theta = \frac{1-\sqrt{5}}{2}}\,.$$

(Already the Golden Ratio ϕ lurks ominously!)

So to solve $(L^2 - L - 1)a_n = 0$, we solve $(L - \phi)a_n = 0$ and $(L - \theta)a_n = 0$, and then linearly combine the results to get a general solution.

4. **Almost there!**

$$(L - \phi)b_n = 0 \Leftrightarrow L(b_n) = b_{n+1} = \phi b_n.$$

But $b_{n+1} = \phi b_n$ is just a geometric sequence equation with the general solution

$$b_n = A\phi^n,$$

where A is some constant. Similarly, $(L - \theta)c_n = 0 \Rightarrow c_n = B\theta^n$.

Thus the general solution of

$$(L^2 - L - 1)a_n = 0$$

is of the form $A\phi^n + B\theta^n = a_n$.

5. The desired formula

To get Jamie Williams's answer, we need to use the "initial conditions" $a_1 = a_2 = 1$. (Note that $a_0 = 0$ can be used to simplify the algebra.)

$$a_0 = 0 = A\phi^0 + B\theta^0 \Rightarrow \boxed{A = -B}$$

$$a_1 = 1 = A\phi^1 + B\theta^1 = (A)(\phi - \theta) = A\left(\frac{1+\sqrt{5}}{2} - \frac{1-\sqrt{5}}{2}\right)$$

$$= A(\sqrt{5}) \Rightarrow \boxed{A = \frac{1}{\sqrt{5}}.}$$

Thus

$$\boxed{a_n = \frac{1}{\sqrt{5}}(\phi^n - \theta^n) = \frac{1}{\sqrt{5}}\left[\left(\frac{1+\sqrt{5}}{2}\right)^n - \left(\frac{1-\sqrt{5}}{2}\right)^n\right]}$$

is the general term of the Fibonacci sequence.

That's one way to do it.

As you can see, this method will work on any recursively defined sequence as long as everything is linear—so it won't work on, e.g., $a_{n+2}^2 = a_{n+1}^2 + a_n^2$. But something like $a_{n+2} - 3a_{n+1} + 2a_n = 0$ should be manageable. $a_{n+2} - 2a_{n+1} + a_n = 0$ will pose the same problems as $(D-1)^2 y = 0$ would in differential equation theory. But you can do it.

Note too that the equations need to have constant coefficients in order for the method to work (i.e., $a_{n+1} = na_n$ is trouble).

If all of this is familiar, sorry to have bored you. If unfamiliar, I hope I've been clear.

Anyway, best of luck!
Take care,
Steve Strogatz
Angel Ct.
Trinity College
Cambridge CB2 1TQ
England

When I look through the folder containing our correspondence, it unnerves me to find so little from 1980 to 1989. Just three letters. And yet it was in 1984 that Mr. Joffray's oldest son, Marshall, died. He would have been 27. We never spoke of this. Mr. Joffray certainly never mentioned it in his letters. Though how can I be sure, since I don't have his letters from this period? Could I have possibly thrown such a letter away, like the rest of them?

I never knew Marshall. I can picture him, though. He was two or three years ahead of me. A handsome boy, tall, with wavy hair and a luminous face. I imagine him (but I don't trust my memory) as a hurdler on the track team.

Anyway, Marshall died, and we never discussed it.

What am I thinking? Why *would* Mr. Joffray have written about this? To me? To anyone. What could he possibly say?

But why didn't I write something to *him*? Did I ever send my condolences? How formal that sounds. What I really mean is, did I ever say how sorry and sad I was to hear that Marshall had died? How awful. And yet I don't believe I ever did. I feel so ashamed about that.

$\int \int \int$

Finally comes the third item in the folder, from 1985. Near the end of my time in graduate school, I paid a visit to Mr. Joffray at his home. The occasion was that my high school friend and fellow grad student, Ed Rak, had wanted to visit "Joff," as he called him. So we arranged a day trip together. Following Ed's lead, it was the first time I ever treated Mr. Joffray as a kind of friend, approachable on almost equal terms. Ed certainly acted that way toward him, so why shouldn't I? Joff cooked hamburgers for all of us and seemed so pleased to see us both, delighted that we'd made the drive down from Boston. We spent a few happy hours playing with his programmable calculator, showing him how to explore chaos, which was then the hottest subject in science.

Looking at the graying Xerox copy of that third item in the folder—a news clipping about some research I'd done—I'm confused about the chronology. The note on the clipping is dated Sept. 2, 1985, and closes with "Hope all is well with you and the other Joffs. Maybe we can arrange another pow-wow this fall."

Does that mean we had visited him the prior spring, the spring of 1985? Or even the preceding fall, of 1984? When exactly had Marshall passed away?

Apparently none of this was the slightest bit on my mind when I wrote, "Hope all is well with you and the other Joffs."

But I do see that I addressed the note to Joff, not Mr. Joffray. That was a shift too. From then on, he would always be Joff to me.

Proof on a Place Mat

(March 1989*)*

Mathematicians aren't known for their social skills. There's an old joke in our community:

> *Q:* How can you tell if a mathematician is an extrovert?
>
> *A:* He looks at *your* shoes when he's talking to you.

But what most people don't realize is that math itself is a very social activity. We mathematicians talk to each other incessantly. We bounce ideas off one another, cogitate together, and get stuck on the same problems together. When you're working on something as hard as math, it helps to share it with someone who understands.

Best of all is when we're able to explain things to each other. I love that. And I'm not alone—mathematicians delight in solving someone else's problem for them.

The legendary physicist Richard Feynman personified this impulse. He was always dazzling his colleagues with his clever tricks. In *"Surely You're Joking, Mr. Feynman!"* he tells the story of how he learned to do integrals by a method called differentiating under the integral sign. His high school physics teacher had made him stay after school one day and said, "Feynman, you talk too much and you make too much noise. I know why. You're bored. So I'm going to give you a book. You go up there in the back, in the corner, and study this book, and when you know everything that's in this book, you can talk again." The book was *Advanced Calculus*, by Woods. Among other things, it taught Feynman the powerful technique of differentiating under the integral sign:

> It turns out that's not taught very much in the universities; they don't emphasize it. But I caught on how to use that method, and I used that one damn tool again and again. . . . The result was, when guys at MIT or Princeton had trouble doing a certain integral, it was because they couldn't do it with the standard methods they had learned in school. . . . Then I come along, and try differentiating under the integral sign, and often it worked. So I got a great reputation for doing integrals, only because my box of tools was different from everybody else's. . . .

Years later, when Feynman was working on the Manhattan Project, he was approached by a colleague whose team had been stuck on a problem for three months. "Why don't you do it by differentiating under the integral sign?" asked Feynman. Half an hour later, the problem was solved.

$\int \int \int$

My correspondence with Joff took off in March 1989, sparked by a question one of his students raised. We sent each other several breathless letters back and forth, including a proof I scribbled on a paper place mat from a Chinese restaurant and mailed to him complete with soy sauce stains.

For whatever reason, it was at this time that I finally began saving Joff's letters. Looking at the first few of them now, almost 20 years later, I'm smiling—he's in such a philosophical mood, riffing about the state of math education, cheerfully wrestling with his calculator, blushing about not having studied more math in college.

Yet my letters back to him are almost exclusively about math and hardly mention anything happening in my life. This was a crucial juncture—the end of my training, the beginning of my career. Why didn't I tell him that I'd just landed my dream job, an assistant professorship in the math department at MIT?

∫ ∫ ∫

The problem that fascinated us had to do with the infinite series

$$\frac{\sin 1}{1} + \frac{\sin 2}{2} + \frac{\sin 3}{3} + \cdots,$$

or what a mathematician would write as

$$\sum_{k=1}^{\infty} \frac{\sin k}{k}.$$

One of Joff's students, Josh Rapoport, had asked him if this series "converges," meaning that the sum gets closer and closer to some limiting number as you take more and more terms. Or does it "diverge," meaning that it either becomes infinitely large or never approaches any limiting number?

This is the sort of question all students in calculus learn to handle (or dread). You're taught various rules to determine if a given series converges or diverges, and it's easy to lose sight of how amazing all this is, how mind-boggling.

You're being asked to think about an *infinite* number of numbers added together. Infinity, right there on the page in front of you. But with calculus, you can handle it nevertheless.

There was a twist, however, in this particular question. The sin k is unusual. You never see that in a first course in calculus. For one thing, it looks weird to be taking the sine function of whole numbers like 1, 2, 3. Normally, we take sine functions of angles, and those are typically simple fractions of π (or 180°), like $\pi/2$ or $\pi/4$. Sine of 1 or 2, without any π's in them, looks bizarre.

The more serious twist, though, is that the sine function oscillates. It can be positive or negative. It's a wave, bobbling up and down. This means that some of the terms in the series will be positive (like $\frac{\sin 1}{1}$ and $\frac{\sin 2}{2}$), while others will be negative (like $\frac{\sin 4}{4}$ and $\frac{\sin 5}{5}$). With plus terms added to minus terms, it seems like the cancellations will help the series to converge.

Joff and all his students would have been thoroughly familiar with a related, and much simpler, series where this kind of cancellation occurs. If the plus signs and minus signs strictly alternate, like this,

$$+-+-+-,$$

and if you replace the sin k terms with $+1$ or -1, you get what's called the "alternating harmonic series"

$$1-\frac{1}{2}+\frac{1}{3}-\frac{1}{4}+\frac{1}{5}-\frac{1}{6}+\cdots.$$

This converges. Here's the argument—it's cute and easy, so try to follow it even if you haven't had calculus. It'll give you a feeling for how we subdue infinity.

∫ ∫ ∫

Imagine that you're walking on the number line, starting from 0. Mark that point L_1. This is your first, and leftmost, point on a journey of infinitely many steps.

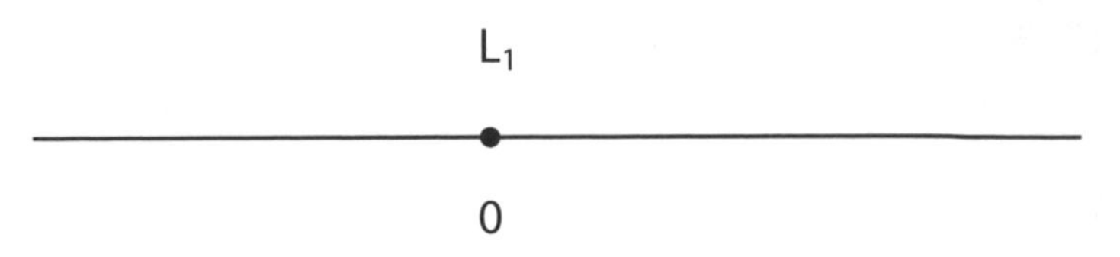

Now walk one mile to the right. Mark the new point R_1, the rightmost point of the journey.

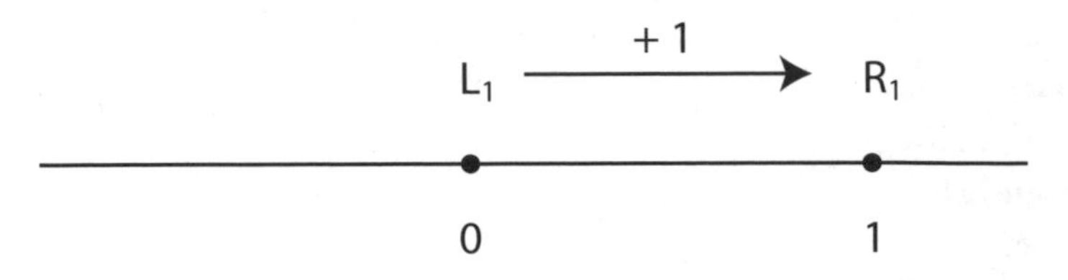

Turn around and walk half a mile back. This represents the first two terms in the series: $1-\frac{1}{2}$. Mark this new point L_2 because it's the leftmost point on this part of your travels.

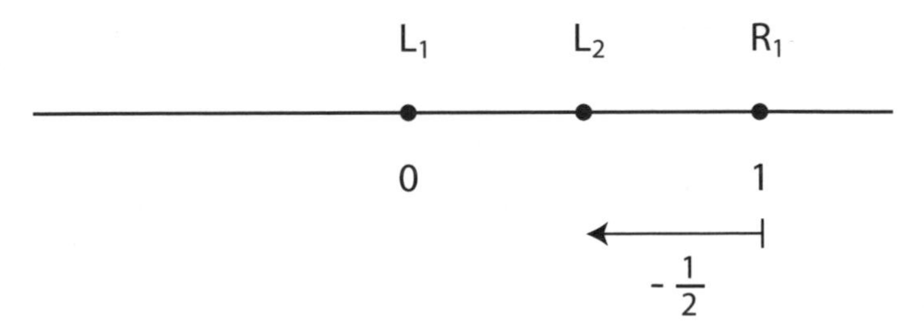

You get the idea. We're going to keep doing this. So turn around again and walk to the right a third of a mile and mark that point as R_2.

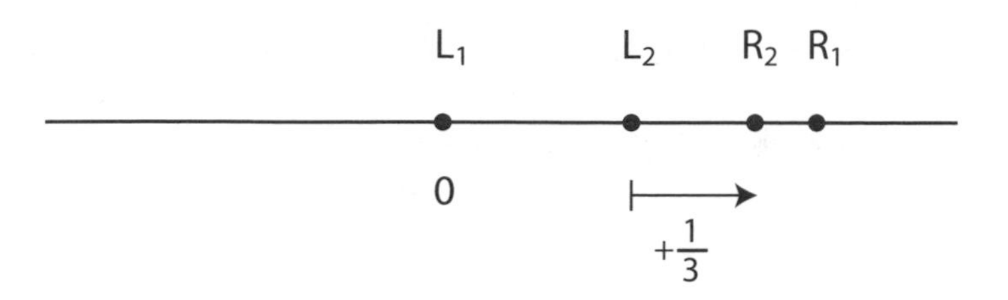

If you want to be numerical about it, $L_2 = 1 - \frac{1}{2} = \frac{1}{2}$ and $R_2 = 1 - \frac{1}{2} + \frac{1}{3} = \frac{5}{6}$. But it's better to think pictorially. Can you see what's happening as we keep repeating this process? It's as if a vise is tightening. The points on the left side of the closing vise are $L_1, L_2, \ldots$, and those on the right are $R_1, R_2, \ldots$.

Each time you turn around on the left, you haven't made it all the way back to the *last* place you turned around on the left. Likewise on the right. Put another way, the interval between consecutive *L*s and *R*s is shrinking. For example, the interval between L_2 and R_2 is nested inside the interval between L_1 and R_1. The same is true for later intervals like L_3 and R_3, though we won't bother drawing them. The point is that the intervals are being squished by the vise, collapsing from both ends. The vise never closes completely, but it gets tighter and tighter. We say the *L*s and the *R*s "converge." Of course, they converge to the same number (since the left and right sides of the vise come closer and closer to touching each other as you add more terms in the series).

What *is* this magic number that the *L*s and *R*s converge to? It's a natural question, but beginners in calculus quickly learn that they're not supposed to ask it. It's usually too hard. As a rule of thumb, it's much easier to prove that a series converges than to find the number it converges to.

Here, for example, our vise argument suffices to prove that $1 - \frac{1}{2} + \frac{1}{3} - \frac{1}{4} + \frac{1}{5} - \cdots$ is some definite number between $\frac{1}{2}$ and $\frac{5}{6}$. We can hem it in as much as we like by taking more terms. Using a computer, we can show that after ten steps we get to

$$L_{10} \approx 0.666$$
$$R_{10} \approx 0.719.$$

After 100 steps,

$$L_{100} \approx 0.6906,$$
$$R_{100} \approx 0.6957.$$

In fact, using more sophisticated ideas of calculus that we needn't go into, one can prove that the limiting number is the natural logarithm of 2, given by $\ln 2 = 0.693147\ldots$.

What does all this have to do with my correspondence with Joff? Well, he was posing the analogous questions for the closely related series $\sum_{k=1}^{\infty} \frac{\sin k}{k}$. Does the series converge? And if so, how can we prove it? He wouldn't have thought to ask what number the series converges to because, remember, that question is taboo in first-year calculus.

We had discussed all these issues by phone on March 6. Joff asked me about $\sum_{k=1}^{\infty} \frac{\sin k}{k}$ and the related integral $\int_1^{\infty} \frac{\sin x}{x}\, dx$. This conversation prompted a barrage of letters. Not only did we revel in the infinite series he asked about but also in Fourier series, Feynman's discussion of differentiating under the integral sign, the gamma function, and a host of related topics. We had a blast.

Dear Steve, Wednesday 3/8/89

It was great to talk with you the other night. I was feeling a might lonely, having prepared my own supper (I cooked hamburgers in the fireplace), and was writing advisee letters when you called. This morn I made an attempt at [a certain math problem], and this didn't look promising.

Maybe I should have integrated under the derivative! Barry Moran and I hope you can send an example of what Feynman referred to in his *Surely You're Joking* book when he dropped "the differentiate under the integral technique."

If you have any light to shed on

$$\int_1^{\infty} \frac{\sin x}{x}\, dx$$

(which I can understand!), I will share it with one of my students who shares my irritation with not being sure whether the integral converges or diverges. I'm wondering if it converges absolutely or needs an infinite sprinkling of negative terms to tip the scale to convergence. Oops! I haven't taken your suggestion to see what a computer look would show. I have a junior in precal/cal who would love to take on that task.

Give my best to Ed.
Thanks for the call Monday eve.
Joff

(Einstein's 110th birthday) March 14, 1989

(Also Telemann's birthday)

Hi Joff,

It was great talking to you the other night. Thanks for your letter too.

As you can see, there are some enclosures with this letter. One is a short note about how to use Love Affairs to "arouse" students' interest in differential equations and their solutions, along with a write-up that appeared in a Chicago paper.

The other enclosure is about the discontinuous function that I call a "finite spike" (to distinguish it from a "δ-function" or infinite spike). That enclosure is somewhat terse, but readable I hope. It is related to a theme of this letter, that is, using an integral or a series to define a function. (You'll see what I mean as we go on.)

First, let's deal with

$$\sum_{k=1}^{\infty} \frac{\sin k}{k}.$$

1. **Integral test fails**

At first I thought of using the integral test because I know $\int_1^\infty \frac{\sin x}{x}\, dx$ is finite. (In fact, $\int_0^\infty \frac{\sin x}{x}\, dx = \frac{\pi}{2}$, as I guessed/remembered on the phone—I'll discuss this integral next. Along the way I'll show you the basics of complex variable theory.)

Unfortunately, even if we did know $\int_1^\infty \frac{\sin x}{x}\, dx$ is finite, we can't use it; the integral test doesn't apply (it applies only if the integrand is positive—sign changes are not allowed). So that didn't work.

2. Alternating series argument—couldn't get it to work but maybe you can

Then I thought of trying the alternating series test, i.e., Leibniz's rule.

[Lots of futile calculations followed.]

3. Computer strongly suggests convergence

Suppose we compute $\sum_{k=1}^{N} \frac{\sin k}{k}$ for different values of N. Using my computer, I found these results in a matter of minutes:

N	$\sum_{k=1}^{N} \frac{\sin k}{k}$
10	1.1245 . . .
100	1.06042 . . .
1,000	1.07069 . . .
10,000	1.070868 . . .
100,000	1.070805 . . .

So we believe that *the sum is close to* 1.0708. . . . Note that I used N *increasing by a constant factor* each time (powers of 10). That's because I know $\sum_{k=1}^{N} \frac{1}{k}$ diverges like $\ln(N)$, and I wanted to check for such logarithmic divergence. If there were any, my partial sums would have grown arithmetically, roughly speaking. Since that didn't happen, I strongly expect $\sum \frac{\sin k}{k}$ converges.

4. OK, it's time to reveal the final word

$$\boxed{\sum_{k=1}^{\infty} \frac{\sin k}{k} = \frac{\pi - 1}{2} \approx 1.070796\ldots}$$

Where does this beautiful result come from? The way I found it involves Fourier series. I will try to explain the main ideas. You and your students may want to look in Thomas's *Calculus* or some other reference for background material on Fourier series.

The usual game in Fourier series is to express some periodic function $f(x)$ in terms of a sum of sines and cosines. Let's discuss only 2π-periodic functions and suppose further that the functions are *odd*. Then we can express $f(x)$ in terms of an appropriately weighted sum of sine functions: $f(x) = \sum_{k=1}^{\infty} a_k \sin kx$, where the a_k are coefficients that depend on the function $f(x)$.

Let's do an example. Consider the function

$$f(x) = \begin{cases} 1 & 0 < x < \pi \\ -1, & -\pi < x < 0 \end{cases}$$

and then periodically repeated for all real x. (At multiples of π, let's set $f(x) = 0$, although it doesn't matter what we do at these discontinuities.) It's a square wave.

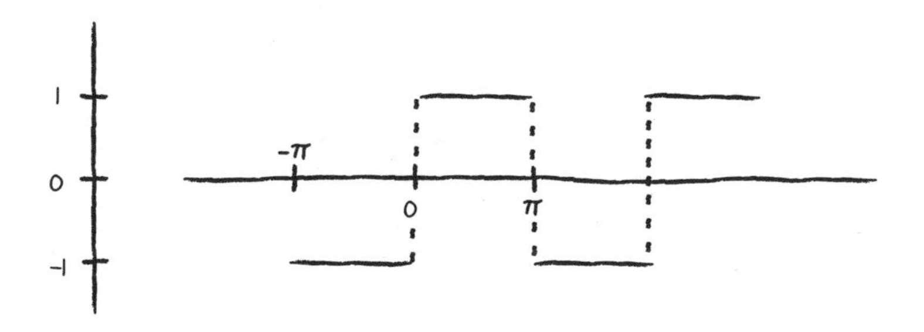

Student Project: Maybe you can hear one of these on Loomis's electronic music synthesizers—I took a course from Mr. Pratt where these things were possible. Square waves sound like oboes, as I recall. They have lots of overtones, which is what we're about to calculate: the strength of each harmonic or overtone. In

$$f(x) = \sum_{k=1}^{\infty} a_k \sin kx,$$

the a_k are the amplitudes of the various harmonics. Let's calculate a_1. It looks like it should be big and positive because $f(x)$ is like $C \sin x$ with squared-off top and bottom:

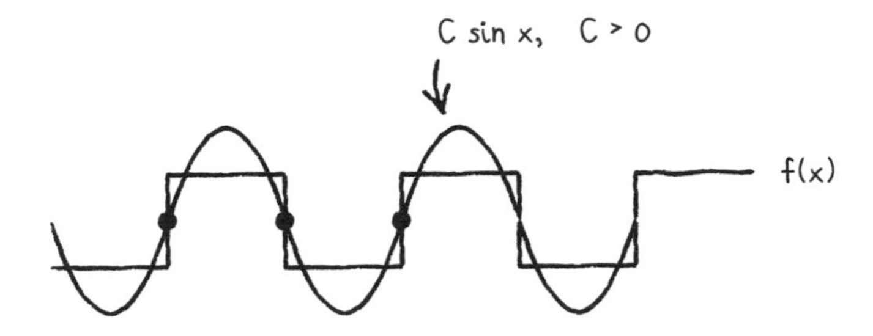

If $f(x) = a_1 \sin x + a_2 \sin 2x + a_3 \sin 3x + \cdots$, then we find a_1 by the following trick: multiply by $\sin x$ and integrate over one cycle. This kills everything except the $a_1 \sin x$ term, which allows us to find a_1. Specifically,

$$\begin{aligned}
f(x) \sin x &= a_1 \sin^2 x + a_2 \sin 2x \sin x \\
&\quad + a_3 \sin 3x \sin x + \cdots \\
\Rightarrow \int_{-\pi}^{\pi} f(x) \sin x \, dx &= a_1 \int_{-\pi}^{\pi} \sin^2 x \, dx + a_2 \int_{-\pi}^{\pi} \sin 2x \sin x \, dx \\
&\quad + a_3 \int_{-\pi}^{\pi} \sin 3x \sin x \, dx + \cdots \\
&= a_1 \pi + 0 + 0 + \cdots \\
&= \pi a_1.
\end{aligned}$$

Notice that all terms after the first integrate to 0; that's the beauty of this.

Hence $a_1 = \frac{1}{\pi}\int_{-\pi}^{\pi} f(x)\sin x\, dx$.

So far we've used nothing about $f(x)$. Now we use our particular $f(x)$ to find the appropriate a_1 for our case:

$$\begin{aligned} a_1 &= \frac{1}{\pi}\int_{-\pi}^{0}(-1)\sin x\, dx + \frac{1}{\pi}\int_{0}^{\pi}(1)\sin x\, dx \\ &= \frac{2}{\pi}\int_{0}^{\pi}\sin x\, dx = -\frac{2}{\pi}\cos x\Big|_0^{\pi} = \boxed{\frac{4}{\pi} = a_1}. \end{aligned}$$

This confirms our intuition that a_1 should be big and positive.

More generally, let's find the kth amplitude: (Apply the same reasoning as before but now multiply by $\sin kx$ to pick off the a_k term.)

$$\begin{aligned} a_k &= \frac{1}{\pi}\int_{-\pi}^{\pi} f(x)\sin kx\, dx \\ &= \frac{2}{\pi}\int_{0}^{\pi}\sin kx\, dx \\ &= \frac{2}{\pi}\left(\frac{-\cos kx}{k}\Big|_0^{\pi}\right) = \frac{2}{\pi k}(1-\cos k\pi) \end{aligned}$$

$$\cos k\pi = \left\{\begin{array}{ll} 1 & k \text{ even} \\ -1 & k \text{ odd} \end{array}\right\} = (-1)^k$$

$$\Rightarrow a_k = \frac{2}{\pi k}[1-(-1)^k]$$

$$\Rightarrow a_k = \begin{cases} \frac{4}{\pi}\frac{1}{k} & k \text{ odd}, \\ 0 & k \text{ even}. \end{cases}$$

(Look at the nice factor of $\frac{1}{k}$ above.)

So our square wave is given by

$$f(x) = \frac{4}{\pi}\sum_{k \text{ odd}}\frac{\sin kx}{k}.$$

Student Project: This would be a nice computer project: plot each of these harmonics and keep adding them up. You'll see a square wave emerging!

Now you're starting to see the connection to $\sum_{k=1}^{\infty} \frac{\sin k}{k}$. Our result $f(x) = \frac{4}{\pi} \sum_{k \text{ odd}} \frac{\sin kx}{k}$ is very close, except (1) we only want to look at the point $x = 1$, and (2) we're missing the even terms in k. When $x = 1$, $f(x) = 1$ (recall $f(x) = 1$ if $x \in (0, \pi)$). Hence, if we believe our Fourier series really represents $f(x)$, we have to conclude that

$$f(1) = 1 = \frac{4}{\pi} \sum_{k \text{ odd}} \frac{\sin k}{k}; \qquad \text{i.e.,} \quad \boxed{\sum_{k \text{ odd}} \frac{\sin k}{k} = \frac{\pi}{4}}.$$

Not quite what we want, but still beautiful. While we're at it, let's put in $x = \frac{\pi}{2}$ for fun. Again $f(x) = 1$ since $0 < \frac{\pi}{2} < \pi$, but now we get

$$\begin{aligned} 1 &= \frac{4}{\pi}\left(\sin\frac{\pi}{2} + \frac{1}{3}\sin\frac{3\pi}{2} + \frac{1}{5}\sin\frac{5\pi}{2} + \cdots\right) \\ &= \frac{4}{\pi}\left(1 - \frac{1}{3} + \frac{1}{5} - \cdots\right) \Rightarrow \boxed{\frac{\pi}{4} = 1 - \frac{1}{3} + \frac{1}{5} - \cdots}, \end{aligned}$$

which you can also get by expanding $\tan^{-1} x$ in a Maclaurin series for $x = 1$.

I hope you see the strategy; we want to think of $\sum_{k=1}^{\infty} \frac{\sin k}{k}$ as a particular value of $f(x)$ for some $f(x)$. That is, we want to write $f(x) = \sum_{k=1}^{\infty} \frac{1}{k} \sin kx$, then plug in $x = 1$ and get $f(1) = \sum_{k=1}^{\infty} \frac{1}{k} \sin k$. But the problem is, what $f(x)$ is being represented by the Fourier series $\sum_{k=1}^{\infty} \frac{\sin kx}{k}$? [This is the "inverse problem"—usually we're given $f(x)$ and asked to find the Fourier series. Here we're asked to find $f(x)$ *given* the Fourier series.]

I did this by going to a table of Fourier series in the math library. It turns out that

$$\sum_{k=1}^{\infty} \frac{\sin kx}{k} = \frac{\pi - x}{2}$$

for $0 < x < \pi$. The picture of the odd periodic function is

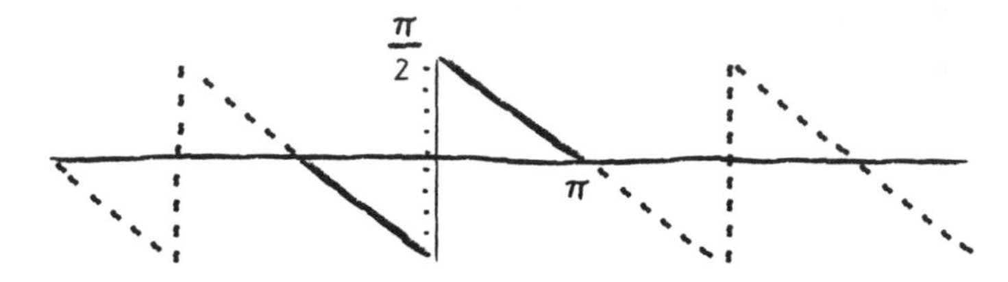

which is a kind of "sawtooth" function. Project: Do it by summing sine waves on the computer or oscilloscope.

Given this, the result

$$\sum_{k=1}^{\infty} \frac{\sin k}{k} = \frac{\pi - 1}{2}$$

follows, by looking at $x = 1$.

This can be seen by mere mortals without having to believe tables of Fourier series. As an *exercise*, you should compute the Fourier series for $f(x) = x$ for $x \in (-\pi, \pi)$.

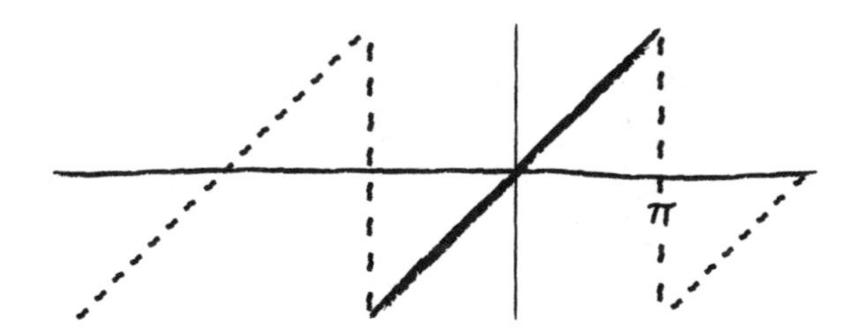

You should get

$$x = 2\sum_{k=1}^{\infty} (-1)^{k+1} \frac{\sin kx}{k}.$$

Plugging in $x = 1$ gives

$$1 = 2\left[\sin 1 - \frac{1}{2}\sin 2 + \frac{1}{3}\sin 3 - \frac{1}{4}\sin 4 + \cdots\right].$$

Combining this with $\sum_{k \text{ odd}} \frac{\sin k}{k} = \frac{\pi}{4}$, as found earlier, we get the missing even terms in k : $\sum_{k \text{ even}} \frac{\sin k}{k} = \frac{\pi}{4} - \frac{1}{2}$. Adding $\sum_{\text{even}} + \sum_{\text{odd}}$, we get $\sum_{k=1}^{\infty} \frac{\sin k}{k} = \frac{\pi}{2} - \frac{1}{2}$.

I'm too tired to show you

$$\int_0^\infty \frac{\sin x}{x}\, dx = \frac{\pi}{2}.$$

But I will show you one "differentiating under the integral sign" trick. It's the best I could come up with but not really what I want to show you. (See the enclosed chapter from *Advanced Calculus* by Woods, the book Feynman used.)

Do you know how to define $n!$ if n is not an integer? This is Euler's idea for interpolating the factorial function—it's called the "gamma function" $\Gamma(x)$.

Suppose n is an integer. Let's show

$$n! = \int_0^\infty x^n e^{-x}\, dx.$$

We could do this by induction, which is why this example isn't so great. Let's do it by differentiating under the integral sign. It saves us the labor of doing lots of integrating by parts. Start with an easy one:

$$\int_0^\infty e^{-ax}\, dx = -\frac{1}{a}e^{-ax}\Big|_0^\infty = \frac{1}{a},$$

where $a > 0$ is a parameter.

Now differentiate both sides of this formula. The parameter a is the thing we're going to be differentiating with respect to. Differentiating under the integral on the left side gives

$$\frac{d}{da}\int_0^\infty e^{-ax}\,dx = \int_0^\infty \frac{d}{da}(e^{-ax})\,dx = \int_0^\infty -xe^{-ax}\,dx,$$

whereas differentiating on the right hand side gives

$$\frac{d}{da}\left(\frac{1}{a}\right) = -\frac{1}{a^2}.$$

Hence

$$\frac{1}{a^2} = \int_0^\infty xe^{-ax}\,dx.$$

Keep playing this game.

$$a^{-2} = \int_0^\infty xe^{-ax}\,dx.$$

$$\text{Differentiate with respect to } a \Rightarrow -2a^{-3} = \int_0^\infty -x^2e^{-ax}\,dx$$

$$\Rightarrow 2a^{-3} = \int_0^\infty x^2e^{-ax}\,dx$$

$$\frac{d}{da}\text{ both sides} \Rightarrow 6a^{-4} = \int_0^\infty x^3e^{-ax}\,dx$$

$$\vdots$$

$$n!\,a^{-(n+1)} = \int_0^\infty x^ne^{-ax}\,dx.$$

Now set $a = 1$. This gives

$$n! = \int_0^\infty x^ne^{-x}\,dx,$$

an "integral representation" of the factorial function $n!$.

The point is that the right-hand side is defined for all real $n > 0$. For example, try to find $\left(\frac{1}{2}\right)!$. You will need to be able to do the definite integral $\int_0^\infty e^{-y^2}\,dy$.

Let me know what you think!

Bye for now,
Steve

I was so jazzed by my correspondence with Joff that I had to share it with someone. The victim was my pal Rennie Mirollo. Rennie and I had been friends for about ten years, having met in 1979 while we were both teaching at the Hampshire summer math program. A few years later we continued our friendship as grad students, and now we were both on the verge of starting our careers as professors.

One day, over lunch at Cheng Feng in Somerville, I mentioned the infinite series problem Joff had posed. I was eager to show Rennie how I'd solved it with Fourier series, but before I could get started, he said there should be a quicker way with complex variables. In between Kung Pao chicken and sesame beef, we worked out Rennie's idea on a red paper place mat.

The idea of the calculation was to use Taylor series, not Fourier series. Recalling Euler's formula $e^{ik} = \cos k + i \sin k$, Rennie realized that the desired sum

$$\sum_{k=1}^{\infty} \frac{\sin k}{k}$$

is just the imaginary part of

$$\sum_{k=1}^{\infty} \frac{e^{ik}}{k},$$

and he knew how to do *that* sum by pattern recognition. Specifically, he spotted that $\sum_{k=1}^{\infty} \frac{e^{ik}}{k}$ is a special case of

$$\sum_{k=1}^{\infty} \frac{z^k}{k}$$

if we set $z = e^i$. And *this* sum is, in turn, a close relative of a famous sum

$$\begin{aligned} z - \frac{z^2}{2} + \frac{z^3}{3} - \frac{z^4}{4} + \cdots &= \sum_{k=1}^{\infty} (-1)^{k+1} \frac{z^k}{k} \\ &= \ln(1+z), \end{aligned}$$

which both of us had seen in courses on complex analysis. Replacing z by $-z$ gives

$$\begin{aligned} -z - \frac{z^2}{2} - \frac{z^3}{3} - \frac{z^4}{4} - \cdots &= -\sum_{k=1}^{\infty} \frac{z^k}{k} \\ &= \ln(1-z) \end{aligned}$$

and hence

$$\sum_{k=1}^{\infty} \frac{z^k}{k} = -\ln(1-z).$$

Next comes a technicality. The series $\sum_{k=1}^{\infty} \frac{z^n}{k}$ is known to agree with $-\ln(1-z)$ for all $|z| < 1$; the question is whether the series *continues* to work if $z = e^i$, which has $|z| = 1$, not $|z| < 1$. In other words, can we use the series *on* the circle of convergence, not just inside it? That's tricky business. Sometimes you can get away with pushing a series out to its "radius of convergence" while other times you get burned. But in problems like this, when you get burned you know it right away because the result is obvious nonsense. So we thought, let's try it and see.

Plugging $z = e^i$ into the logarithm and taking the imaginary part gave $\sum_{k=1}^{\infty} \frac{\sin k}{k} =$ imaginary part of $-\ln(1 - e^i)$. Next, to evaluate this imaginary part, we set

$$1 - e^i = re^{i\theta}$$

so that

$$\begin{aligned} -\ln(1 - e^i) &= -\ln(re^{i\theta}) \\ &= -\ln r - i\theta. \end{aligned}$$

Thus, the imaginary part is $-\theta$, and the problem has been reduced to figuring out what θ equals, given that $re^{i\theta} = 1 - e^i$.

That final step was an exercise in trigonometry. Here's one way to solve for θ. Taking the real parts of both sides of $1 - e^i = re^{i\theta}$ gives $1 - \cos 1 = r \cos\theta$. Likewise, the imaginary parts give $-\sin 1 = r \sin\theta$. Now divide the second equation by the first. The r's cancel, and we get

$$\tan\theta = \frac{-\sin 1}{1 - \cos 1}.$$

The right hand side can be simplified by applying the half-angle identity

$$\cot(x/2) = \frac{\sin x}{1 - \cos x}$$

to the case $x = 1$. Thus

$$\frac{-\sin 1}{1 - \cos 1} = -\cot(1/2)$$

and hence

$$-\theta = \tan^{-1}\left(\cot \frac{1}{2}\right).$$

At this stage Rennie drew a little triangle

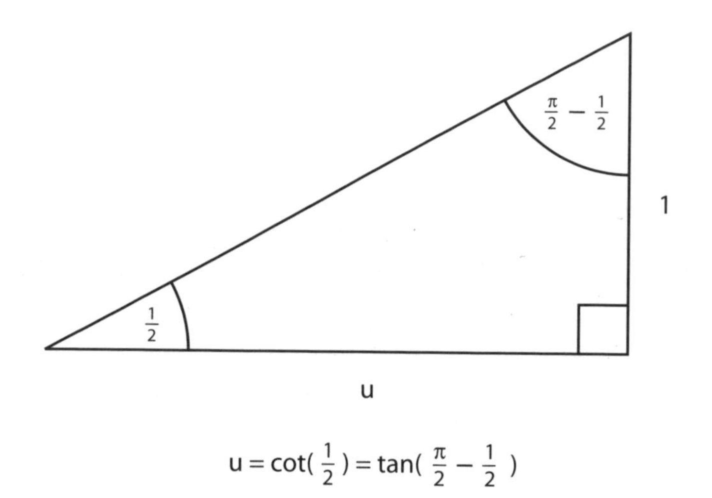

$$u = \cot\left(\frac{1}{2}\right) = \tan\left(\frac{\pi}{2} - \frac{1}{2}\right)$$

which demonstrated that

$$\tan^{-1}\left(\cot \frac{1}{2}\right) = \frac{\pi}{2} - \frac{1}{2},$$

and that was the answer we were looking for.

Joff loved this whole argument: the use of complex numbers, the little tactics with the triangle, and the nonchalant assumption that the series was inside its radius of convergence. In the years to come, he'd often show the place mat, like some sacred scroll, to his latest crop of advanced students.

Monday eve.

Dear Steve, March 20, 1989

It is late in the day that you sent me the Mirollo solution to $\sum \frac{\sin k}{k} = \frac{\pi - 1}{2}$, and I have put aside all other projects to work through the steps. I am grinning like a Cheshire cat, since it all makes sense (with someone else leading the way) and is so nice. I must say I appreciate seeing the *use* of things that as a high school teacher I put on the board: (complex # equality, polar/rectangular equivalents . . .).

* You just called on the phone, so there isn't much to say at the moment except more thanks—this time for giving me new insights on the use of $\sum \frac{\sin k}{k}$ in diffraction, the analogies between vectors and Fourier series, and a host of other notions.

I'm speechless about how you and Rennie came up with $\tan^{-1}(\cot \frac{1}{2})$. I have a new appreciation for the use of the appropriate right triangle for working composition of trig and inverse trig functions.

I have just written myself a note to pursue the road not taken in your collaboration with Rennie to show me $\sum \frac{\sin k}{k}$. This may be one of the best spring breaks I've ever had at Loomis, thanks to your correspondence and phone calls!

** I decided to take a break and work on $\sum \frac{\cos k}{k}$. My mathematics gave me $\sum_{k=1}^{\infty} \frac{\cos k}{k} = -\ln(2 \sin \frac{1}{2}) \approx$ .0420195. As a check I ran a program on my TI58 and got $\sum_{k=1}^{2450} \frac{\cos k}{k} \approx .042065$. I noted that my program step counter had accumulated some error in incrementing by 1 over and over again. At the moment $k = 2714.$**08452** and the sum is . . . oops! I must have fouled up the memory that stored the sum; it is now .08. Ugh. It is also late, so I'll turn in and revisit this when I'm rested!

Tuesday

During breakfast I reran my program (with no end command). When I finally broke in on it, $k = 6719$ (no decimal part this time!) and $\sum \frac{\cos k}{k} = .0421715$. Maybe one of us (me or the TI58) is getting too old. I remember you using this calculator in the homestead backyard while Ed and you visited. As I cooked hamburgers, you demonstrated some chaos on this calculator. That was quite a while ago. Maybe my calculator has been secretly generating its own personality with the remnants of what you left on it.

What's left of my sanity thinks that a well-behaved TI58 should be able to count from 1 to 2714 using binary and not come out with 2714.08452. The other glitch of last night seems even more unaccountable. At least today the thing can count to 6719 with no visible decimal part to confuse me. The overestimate of what I hope $\sum \frac{\cos k}{k}$ adds up to suggests that convergence wanders above and below the limiting number as $k \to \infty$. You have already addressed a similar problem in trying to group $\sum \frac{\sin k}{k}$ to alternate so you can use Leibniz for a convergence test.

If my work is correct, $-\ln(2\ \sin\frac{1}{2})$ is strange but attractive in its simplicity. (Is Hofstader feeding my brain now? . . . Didn't he use the term "strange attractor"?)

This morning I took on your task of finding $(\frac{1}{2})!$. Lord Kelvin would not have been too pleased when I thought I remembered $\int_0^\infty e^{-x^2}\,dx = \frac{\pi}{\sqrt{2}}$. Low marks, Donald! Luckily, there is a lad at Harvard who just happened to send along a picture of a text that included the double integral solution to that famous integral. **I** think $(\frac{1}{2})! = \frac{\sqrt{\pi}}{2}$.

My calculations (bare bones steps):

$$\left(\frac{1}{2}\right)! = \int_0^\infty x^{1/2} e^{-x}\, dx = \int_0^\infty 2y^2 e^{-y^2}\, dy$$
$$= -ye^{-y^2}\Big|_0^\infty + \int_0^\infty e^{-y^2}\, dy;$$
$$\left(\frac{1}{2}\right)! = 0 + \frac{\sqrt{\pi}}{2} \approx .886.$$

(To do that y integral, I used integration by parts with $dv = e^{-y^2} 2y\, dy$.) The end result seemed reasonable because of its elegance, but my curiosity then asked for what positive real number n is $n!$ least?

After trying a little differentiation under the integral and coming up short, I tried to check $(\frac{1}{2})!$ on my calculator by calling up the Simpson's rule program in its library. A first run, doing $\int_0^{1000} \sqrt{x}\, e^{-x}\, dx$ using $n = 500$ subdivisions gave me about .577. . . . That was hardly reassuring. I was thinking that there'd be rather quick convergence. Has my TI58 rebelled? It *couldn't* be my programming ineptitude! Could my calculus be flawed? Should my teaching license be revoked or at least suspended? Arghh! @%X!

While I'm waiting for my program to run (I'm trying *once* more), I meant to ask if you have ever read *The Adolescence of P-1* by Ryan? We have a copy in the Loomis library. I thought it was a light and entertaining novel written about a young college lad who had a regenerative program get away from him. You might enjoy it. It was written about 10 years ago, I think. There was mention of Martin Gardner's article in *Sci. Am.* on "how to teach a match box to play tic-tac-toe." —Early days of AI stuff.

My TI58 is playing someone else's tune, not mine. I grabbed my CASIO FX 7000 G manual and attempted to input their canned Simpson's rule program. The Casio then insulted me with a collection of GO and SYN errors. I am going to take a walk with my beagle and maybe fashion an abacus from sticks and dried wild cucumbers.

When my frustrations have subsided, I will return with renewed vigor to explore the Strogatzian adventures you have mapped for me.

In the meantime I send you my admiration and gratefulness—

and best regards,
Joff

Dear Steve, Wed. eve. March 22, 1989

Since I have written you twice at Sue's office I have inadvertently sent some of my correspondence to H22 Winthrop. One letter and one postcard, I think. I hope they find you (or are you maintaining double residence in Winthrop and Pierce?).

We had in-service day for the faculty today. This is an invention of the administration that gets the teachers back a day ahead of the students for the purpose of attending stimulating sessions on some facet of education. Today the subject was "Writing Across the Curriculum." I must say that Sheldon Glashow's talk at the W.A.L.K.S. physics meeting at Kingswood last fall made today's content look like . . . antisuperlatives fail me! While we were being lectured that the teaching of writing is a responsibility of all departments (for the nth time, $n \to \infty$), I wrote a lot of mathematics, most of it inspired by your recent instruction.

In working out the Fourier tasks you set for me, I had to convince myself that the zapping device used to evaluate the coefficients really worked. It may be late in my career for such a relevation, but I needed what I have come to call the camel hump theorem:

$$\int_{-\pi}^{\pi} \sin^2 kx \, dx, \qquad k \in \mathbf{N}, = \pi, \text{ no matter what } k.$$

Geometrically I saw this as a squeezing of a camel's humps as k escalates through the natural numbers. I *did* know $\int_{-\pi}^{\pi} \sin^2 x \, dx = \pi$ from the graph

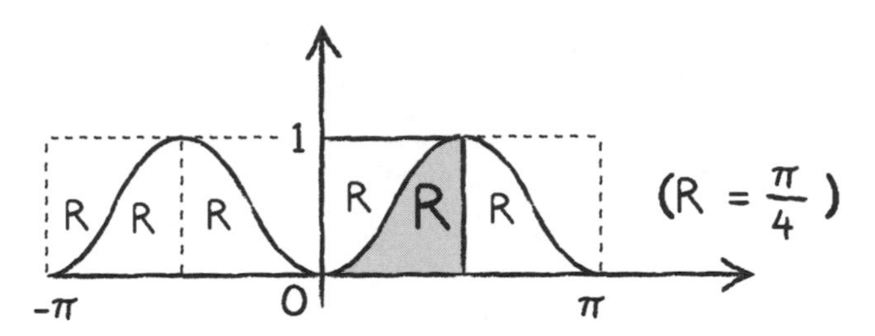

The region R always reminds me of a jigsaw puzzle piece which fits in all the spaces above.

So you have given me a boost by pushing me into the Fourier work. The jigsaw puzzle pieces just get compressed, accordion fashion, and suddenly I'm seeing them as camel's humps. Humph!

Here is my artistry for $k = 3$:

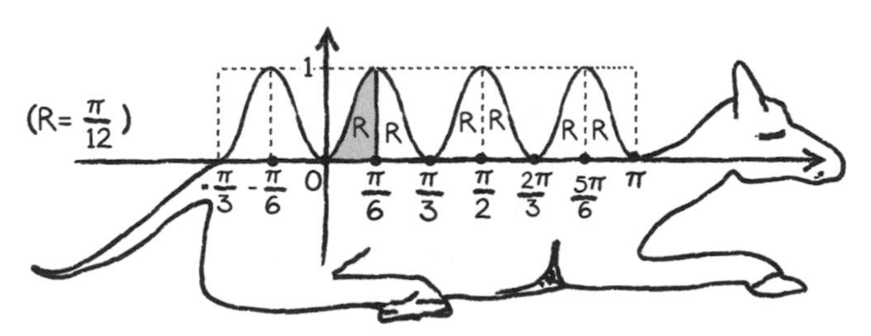

Then I also needed to convince myself that the terms, other than the $\sin^2$ term, got zapped:

$$\int_{-\pi}^{\pi} \sin x \sin kx \, dx = 0 \qquad \text{for} \qquad k \in \mathbf{N}, k \geq 2.$$

I used

$$\begin{array}{r} \cos(x+y) = \cos x \cos y - \sin x \sin y \\ \cos(x-y) = \cos x \cos y + \sin x \sin y \\ \hline \cos(x-y) - \cos(x+y) = 2 \sin x \sin y \end{array}$$

to do the integral as a "sum" of cosines. It reminded me that it's been a long time since I've emphasized the utility of the factoring identities for sines and cosines. Once upon a time I included them in discussions of periodicity. That was when we met 5 times a week and classes were 50 minutes. No wonder we are being shown up by our rival nations in our math scores and abilities. We've cut back to 4 classes/week of 45 minutes. The high schools let out at 2 p.m. Maybe U.S. high school kids are better than the Japanese at bagging groceries in supermarkets (so they can drive to school in racey Pontiacs with quadrophonic stereos).

But I digress!

As I reviewed the place mat proof that $\sum \frac{\sin k}{k} = \frac{\pi - 1}{2}$ and got to the $\tan^{-1}(\cot \frac{1}{2})$ step (this in the middle of our in-service lecture), a light flashed and I saw it as $\tan^{-1}\left[\tan\left(\frac{\pi}{2} - \frac{1}{2}\right)\right] = \frac{\pi}{2} - \frac{1}{2}$. I include this in case my post-card to you gets lost in the mail.

Classes tomorrow, I meet my Calculus AB seniors. I'm afraid that too many of them have been pushed to accelerate in the 8th grade and aren't sharp enough (or gung ho about math enough) to retain the basic ingredients one ought to enter the study of calculus with. Many have forgotten double angle forms for sine and cosine, periodicity, algebra, . . . we haven't come to log and exp topics yet. I cringe to think of how much review this will need to get us to the calculus of ln x, e^x.

I showed your diff. under the integral to Barry Moran. He's working on $(\frac{1}{2})!$. For goodness sake don't feel you have to answer my barrage of correspondence. I just want you to know that all your efforts aren't sitting on a cluttered desktop. I'd read about spike functions in my son Jeff's Clarkson math/eng text which I'm afraid has been stolen.

More thanks + best regards,
Joff

Dear Joff, March 28, 1989

Thanks for your postcards and letters. I've gotten them all—Winthrop is my home, and Pierce is my office.
I loved the camel hump! Your artwork is as nice as I remembered it! Here's a point you may have noticed, but I'll mention it anyway. The complementary part of your camel jigsaw is $\int_0^\pi \cos^2 kx\, dx$, turned upside down:

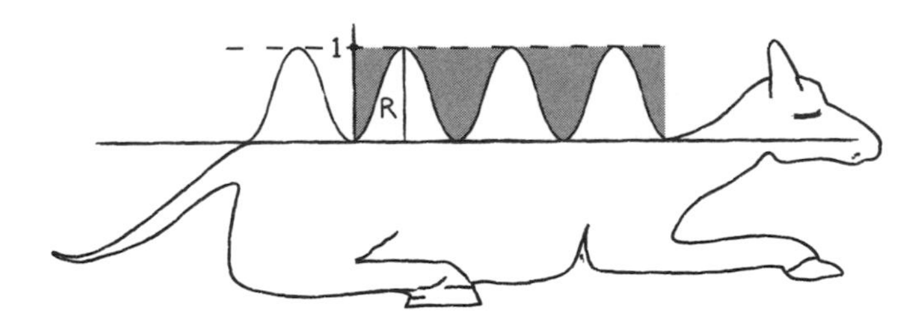

Now since $\cos^2 kx + \sin^2 kx = 1$,

$$\int_0^\pi \cos^2 kx\, dx + \int_0^\pi \sin^2 kx\, dx = \int_0^\pi dx = \pi,$$

and since the two integrals on the left hand side are equal, we conclude that $\int_0^\pi \sin^2 kx = \frac{\pi}{2}$. This is equivalent to your nice picture with rectangles. Professionals remember this by saying that "the average of $\sin^2$ over a cycle is $\frac{1}{2}$."

Your work on $\sum \frac{\cos k}{k}$ agreed with mine. What great stuff!

As for the $\tan^{-1}(\cot \frac{1}{2}) = \frac{\pi}{2} - \frac{1}{2}$, another nice picture is this one:

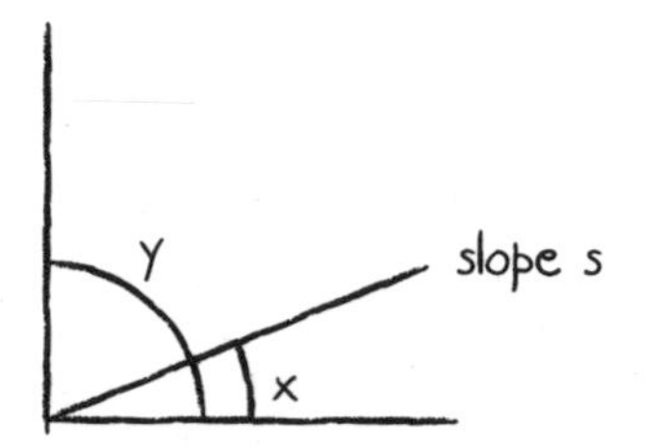

s = tan x = cot y
where $x = \frac{\pi}{2} - y$

I also enclose a graph of the Γ function, $\Gamma(x) = (x-1)!$, for real x. Apparently the minimum is near $x = 1.46\ldots$ (I don't see why yet.) So (.46)! is the smallest for $x > 0$. Amazing! ← *Not* factorial symbol. Say hi to Barry Moran for me.

Cheers,
Steve

The Monk and the Mountain

(1989-90)

In the June 1961 Mathematical Games column in *Scientific American*, Martin Gardner posed a riddle that has become a favorite in courses on the psychology of creativity:

> One morning, exactly at sunrise, a Buddhist monk began to climb a tall mountain. The narrow path, no more than a foot or two wide, spiraled around the mountain to a glittering temple at the summit.
>
> The monk ascended the path at varying rates of speed, stopping many times along the way to rest and to eat the dried fruit he carried with him. He reached the temple shortly before sunset. After several days of fasting and meditation he began his journey back along the same path, starting at sunrise and again walking at variable speeds with many pauses along the way. His average speed descending was, of course, greater than his average climbing speed.
>
> Prove that there is a spot along the path that the monk will occupy on both trips at precisely the same time of day.

You might want to think about this for a minute. Can you come up with a convincing argument why this has to be true? You don't have to figure out where the magic spot is—its location will depend on the unknown details of how fast the monk walked, how long he paused, and so on. You just have to prove that such a spot must exist, somewhere.

Many people feel that this problem is impossible, that not enough information has been given. And it's certainly true that you can't solve it with words or algebra.

Give it a try and then read on.

One approach is pictorial. Suppose we draw a graph of where the monk is at different times of day. On the ascent, the graph starts at the bottom at sunrise, then rises on a wiggly path (wiggly because of the monk's pauses and variable walking speed), and finally reaches the top just before sunset.

Now think about what the corresponding graph looks like on the way down. It's another wiggly curve. We know almost nothing about its shape, except that it starts at the top and ends at the bottom.

Fortunately, that's enough information to solve the problem. When you put the rising and falling graphs on the same axes, it's clear they have to cross each other somewhere. At that intersection point, the monk is at the same place at the same time on both days.

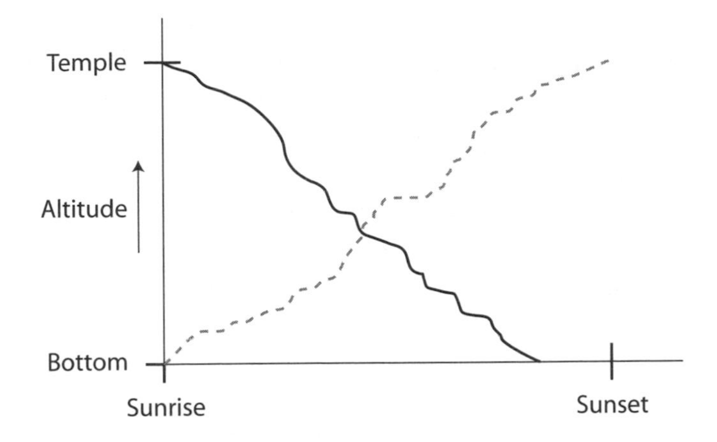

If you don't find this argument convincing enough, you can make it more rigorous with calculus. Assuming that both curves are continuous functions of time (meaning that the monk moves continuously along the path and does not jump from point to point or use a jet pack), the intermediate value theorem guarantees that the rising and falling curves have to cross somewhere.

All this is fine, but it's still not the best argument. The best argument uses only common sense and some imagination. It's a nice example of the role of visualization in a mathematical proof.

Imagine the monk as *two* people, one climbing the mountain, one descending, and both starting at dawn on the same day. In other words, superimpose the events of both days and watch what happens. His climbing self will encounter his descending self somewhere on the mountain. And when he does, he'll be at the same place at the same time. QED.

∫ ∫ ∫

When I look back on where Joff and I were in the fall of 1989, I'm reminded of the monk and the mountain. Our correspondence was about to take off like never before, right at the point where our careers began to cross, his descending and mine on the rise. We were now at the same place at the same time, though taking separate journeys.

It was a happy time for both of us. Joff's letter of May 7, 1990, describes the house he'd later retire to:

> Sue and I have recently purchased a home in Old Lyme, CT, with a magnificent view of a salt marsh. Yesterday we watched a pair of bluebirds trying to decide which of my nest boxes to use, later a red fox strolled through the marsh. At dusk, while we ate dinner, a doe with her fawn entertained us by frolicking in the marsh.

Meanwhile I'd just begun my job at MIT. I was ecstatic. Joff and I were now fellow teachers. We both had students, and we could share the calculus problems we were giving them.

Plus I'd finally fallen in love. Elisabeth was an engineer, an aerobics instructor, and Jewish too—about as close to my fantasy woman as possible. She was sweet and fun and she loved me back. We seemed to be headed for marriage.

Time to meet the parents.

"Doesn't Elisabeth talk?" was my mother's question to me afterward. Well, yes, she may have been a bit nervous or intimidated, but overall, things went well, and everyone was happy, more or less. I was aware of a whiff of disapproval on my mother's part, but it was just that, a whiff, and later encounters would go better, no doubt.

∫ ∫ ∫

During one of our phone calls, Joff mentioned an unusual problem that one of the science teachers at the school had consulted him about. It had to do with a "nonlinear oscillator." The normal thing to study is a linear oscillator, more commonly known as a simple harmonic oscillator, where a mass hangs from a spring and bobs up and down like a daredevil hanging from a bungee cord. The simplest assumption is that the spring obeys a linear force law: the more you stretch the spring, the more it pulls back, in direct proportion to the stretch. In that case, you can prove that the time taken for one complete vibration is always the same, no matter how big or small the vibration is. In other words, the period of the oscillation is independent of its amplitude. This is something that all high school physics students learn and that their teachers feel secure about.

But what if the force law were not linear? For instance, what if it were cubic, $F = -kx^3$, meaning that a stretched spring pulls back with a force proportional to the cube of its stretch? Would the period of vibration still be independent of amplitude? Or if not, would it increase or decrease for larger amplitudes?

My answer to Joff uses a method known as "dimensional analysis." It gives you a quick, back-of-the-envelope approach to solving problems like this. In fact, you don't even need calculus to use it—just algebra and some basic knowledge about the units (the dimensions) of the physical quantities that come into the problem. When a question is too tough to tackle by any other means, dimensional analysis is often worth a shot.

Hi Joff, April 16, 1990

I had some fun thinking about the period of a mass m on a spring with force law $F = -kx^3$. A good approach to such problems, which I forgot to mention, is called "dimensional analysis." You use the *units* of the quantities in the problem to get the essential dependences among variables.

Example: Suppose we say the initial displacement is A and the mass is released from rest. Then there are only three parameters in the problem: A, k, and m.

- A is a displacement—it is measured in meters, or whatever. The point is, it has the dimensions of length. We'll denote this idea by writing $A \sim L$. This means "A has the units of length."
- m of course has the units of mass. So we'll write $m \sim M$.
- What about k? Well, it's defined by $F = -kx^3$. F is a force, which has units of mass × acceleration; i.e., $F \sim ML/T^2$ (it could be expressed in kg-meter/sec^2, for example). x^3 has units $x^3 \sim L^3$. So

$$k \sim \frac{F}{x^3} \sim \frac{ML}{T^2L^3} \sim \frac{M}{L^2T^2}.$$

Summarizing:

$$\left.\begin{array}{l} A \sim L \\ k \sim M/(L^2T^2) \\ m \sim M \end{array}\right\} \leftarrow$$

Since we're looking for a period $\sim T$, we have to combine these, and *only* these, in some way to get a quantity that has the units of *time*.

The M and L must cancel, so that we get a pure time. The only combination of A, k, and m that has units of T is

$$\left(\frac{1}{k} \cdot \frac{1}{A^2} \cdot m\right)^{1/2} \sim \left(\frac{L^2T^2}{M} \cdot \frac{1}{L^2} \cdot M\right)^{1/2} \sim T.$$

In other words, without doing any real work, we know the period must be proportional to $\frac{1}{A}\sqrt{\frac{m}{k}}$!

A *systematic* way to find the appropriate combination of m, k, and A is to write

$$m^{\alpha}k^{\beta}A^{\gamma} \sim T^1 \sim T^1M^0L^0$$

and solve for α, β, γ, like so:

$$\begin{aligned}(m)^{\alpha}(k)^{\beta}(A)^{\gamma} &\sim (M)^{\alpha}\left(\tfrac{M}{L^2T^2}\right)^{\beta}(L)^{\gamma}\\ &\sim M^{\alpha+\beta}L^{-2\beta+\gamma}T^{-2\beta},\end{aligned}$$

and this must give a pure time $M^0L^0T^1$. Hence

$$\begin{aligned}\alpha+\beta&=0\\ -2\beta+\gamma&=0\\ -2\beta&=1\end{aligned}$$

$\Rightarrow \beta=-\frac{1}{2}, \gamma=-1, \alpha=\frac{1}{2};$ i.e., we need

$$\begin{aligned}m^{\alpha}k^{\beta}A^{\gamma} &\sim m^{1/2}k^{-1/2}A^{-1}\\ &= \sqrt{\frac{m}{k}}\,\frac{1}{A}.\end{aligned}$$

(As claimed.)

Warning:

This method is dangerous if you forget other parameters that might be relevant. For example, if we use the same method on a pendulum of length ℓ, mass m, and initial amplitude θ_0, we could write the following:

$$m \sim M$$

$$\ell \sim L$$

$$g \sim \frac{L}{T^2}.$$

This is the thing you might forget $\rightarrow \theta_0 \sim$ dimensionless (it's a number—no units!).

So T *could depend on* θ_0 (*in some unspecified way*) or on some combination of m, ℓ, and g. You can see the *only* way to get a time is to use the combination $\sqrt{\frac{\ell}{g}}$. Notice that the mass m can't possibly enter. There's nothing to cancel it! The trick is that θ_0 could enter (and of course it does—the period *does* depend on θ_0!).

Exercise: (a) Show that (in contrast) the period of a simple harmonic oscillator can't possibly depend on its initial amplitude. (b) What about if an initial velocity $v_0 \neq 0$ is also given? (c) Redo all the calculations above for force laws of the form $F = -kx^n$, where n is odd and positive. What is an expression for the dependence of the period on k, m, and A?

Now let's turn to the integral

$$\int_0^1 \frac{dx}{\sqrt{1-x^4}}$$

that you mentioned came up when you tried to calculate the period of the cubic oscillator by using conservation of energy. As I guessed (and you did too), it's an elliptic integral. I'll explain those some other time. For now let's just see what one can do to get a number out.

First, notice the integral is *finite* since $1-x^2 \leq 1-x^4$ for $0 \leq x \leq 1$, and so

$$\begin{array}{ccc} \displaystyle\int_0^1 \frac{dx}{\sqrt{1-x^2}} & \geq & \displaystyle\int_0^1 \frac{dx}{\sqrt{1-x^4}} \\ \| & & \\ \sin^{-1} 1 & = & \dfrac{\pi}{2}. \end{array}$$

So we know

$$\boxed{\int_0^1 \frac{dx}{\sqrt{1-x^4}} \leq \frac{\pi}{2} \sim 1.57.}$$

Next, I resorted to looking in the mathematical tables by Abramowitz & Stegun (the Bible of such stuff). There in Chapter 17, Elliptic Integrals, I found Formula 17.4.56, which says essentially

$$\frac{1}{\sqrt{2}} K\left(\frac{1}{2}\right) = \int_0^1 \frac{dx}{\sqrt{1-x^4}},$$

where

$$K(m) = \int_0^1 [(1-t^2)(1-mt^2)]^{-1/2}\, dt$$

is called the "complete elliptic integral of the 1st kind." (No one remembers this stuff—"you can always look such things up if you need to" ← a comment I *hate* to hear, but this time it's true.)

The table then tells us that

$$K\left(\frac{1}{2}\right) = 1.8540746773\ldots(+\ 10 \text{ more digits!}).$$

So the prediction is

$$\int_0^1 \frac{dx}{\sqrt{1-x^4}} \approx \frac{1}{\sqrt{2}}(1.854\ldots)$$

$$\approx \boxed{1.31103\ldots}$$

(which *is* $\leq \frac{\pi}{2}$, as expected).

[Next came some computer calculations, omitted here for brevity. I asked the machine to generate the first 40 terms of the Maclaurin series for the integrand $(1 - x^4)^{-1/2}$ and then integrate them term by term. The resulting estimate of the definite integral was 1.22, disappointingly far from the expected answer of 1.31. The culprit was the slow convergence of the Maclaurin series I was using.]

Warmest regards,
Waiting for the next question,

Steve

P.S. 2:15 a.m. (!)

Hey Joff,

Before falling asleep, I thought of a much more rapidly converging series for $\int_0^1 \frac{dx}{\sqrt{1-x^4}}$. The problem in our first try comes from the $\frac{1}{\sqrt{1-x}}$ singularity, so just leave that in there and expand the rest.

The best move may be

$$(1-x^4)^{-1/2} = \underbrace{(1+x^2)^{-1/2}}(1-x^2)^{-1/2}$$

Expand this only (it's not singular near $x=1$)

$$(1+x^2)^{-1/2} = 1-\frac{1}{2}x^2 + \left(-\frac{1}{2}\right)\left(-\frac{3}{2}\right)\frac{1}{2}x^4 - \cdots$$

$$= 1-\frac{1}{2}x^2 + \frac{3}{8}x^4 - \cdots$$

$$\Rightarrow \int_0^1 \frac{dx}{\sqrt{1-x^4}} = \int_0^1 \left(1-\frac{1}{2}x^2 + \frac{3}{8}x^4 - \cdots\right)\frac{dx}{\sqrt{1-x^2}}.$$

The point:

Each integral of the form $\int_0^1 \frac{x^{2n}\,dx}{\sqrt{1-x^2}}$ is done by trig substitution.

Let $x = \sin\theta$.

$$\int_0^1 \frac{x^{2n}}{\sqrt{1-x^2}}\,dx = \int_0^{\pi/2} \sin^{2n}\theta\,d\theta$$

$$= \frac{\pi}{2}\,\frac{1}{2}\cdot\frac{3}{4}\cdot\frac{5}{6}\cdots\frac{2n-1}{2n} \qquad \text{(Wallis formula)}$$

So just use Wallis to your heart's content!

$$\int_0^1 \frac{1-\frac{1}{2}x^2+\frac{3}{8}x^4-\cdots}{\sqrt{1-x^2}}\,dx$$

$$= \frac{\pi}{2}\left[1-\frac{1}{2}\left(\frac{1}{2}\right)+\frac{3}{8}\left(\frac{1}{2}\cdot\frac{3}{4}\right)-\cdots\right]$$

$$\approx \frac{\pi}{2}\left[\frac{64-16+9}{64}\right] = \frac{57}{64}\,\frac{\pi}{2} \approx 1.399.$$

This series alternates, so we always get (improving) upper and lower bounds on the answer.

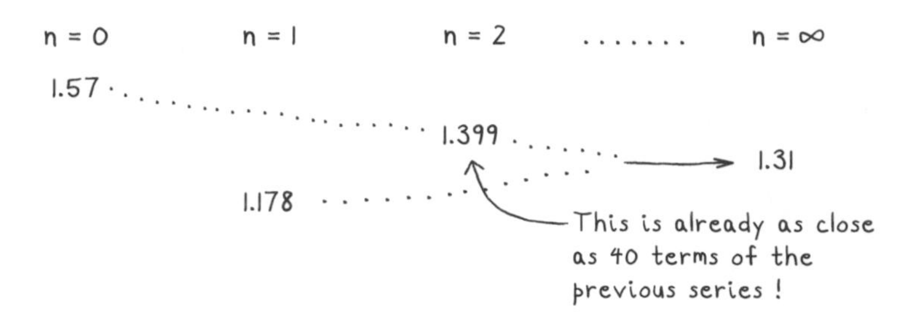

Later that spring Ed Rak and I were invited to return to Loomis to give a speech about Joff at a faculty honors banquet. Joff had won the 1990 Swan Award for Teaching, given by a consortium of schools "to a teacher whose achievement represents the highest standards of the profession."

Ed and I had been asked to prepare a few words and surprise Joff with our presence. We'd become friends in graduate school but been aware of each other long before that, back when we were on the math team in high school. Two years younger than me, Ed had a mathematical mystique. Teachers would talk about him, about how he'd get perfect scores, 18/18, each week at these interscholastic contests. He was likable, not nerdy, and not competitive either. Just matter-of-fact perfect. The only mannerism that comes to mind is that he cleared his throat often, and noisily. A barking sound. But for a math genius, this is a pretty minor mannerism.

Now here we were, back at the school again, in the stately old dining room—chandeliers, candles on the table, big windows overlooking the quad, evening falling. The time came for us to speak. We stood at our round dinner table, faces of faculty, our old teachers, looking up at us, smiling, expectant. I had prepared only a few off-the-cuff remarks and didn't bring any written notes. The start felt okay. But soon enough, in front of everyone, I got so choked up I couldn't continue. Ed was standing right beside me, and he quickly jumped in. Good, strong, affectionate words about Joff. No throat clearing, either.

The rest of it I don't recall. Did I sit down while he spoke? Did I remain standing, trying to compose myself? It's all a blur, really. But I'm glad Ed was there.

Randomness

(1990-91)

It was in the summer of 1990 that a red flag went up for my mother about Elisabeth. Elisabeth felt financially insecure in Boston and suggested that maybe we should move to Troy, New York. I'd been approached by Rensselaer Polytechnic Institute about a job there, the lure being early tenure. Elisabeth liked the idea because she'd grown up in that part of upstate New York and the lifestyle and affordability appealed to her.

Well, that didn't sit well with my mother. Not at all.

Actually, Elisabeth had been rightly unhappy about many things for quite a while. She'd given up a job she loved in Huntsville, Alabama, so we could be together (we'd been having a long-distance relationship for the past year). She'd moved into my tiny apartment, too small for both of us. So small that she had to keep her clothes in storage. Unemployed and adrift, she became increasingly irritable. My mother said, "Of course Elisabeth is unhappy. She wants to get engaged. She's a good girl. Go ahead and propose to her. You want to, don't you?"

Elisabeth and I got engaged.

But when Elisabeth started talking about RPI, leaving Boston and MIT, my mother started to question her own advice. Maybe we should think again about this wedding. When I raised the question with Elisabeth, she became so infuriated that she may have hit me, or thrown something at me, maybe the engagement ring.

We went into couple therapy, and the therapist seemed alarmed about what was going on between us. Still, the wedding plans continued.

Then, early one morning in October 1990, the phone rang and woke me and Elisabeth up. My dad said, "Steven, I have some very bad news. Mom died last night."

He went on to tell me about how her intestine had become twisted around itself. While the doctors were operating on her, she had a heart attack and died on the operating table.

At the funeral I sat with Elisabeth, my dad, and my brother and sister. My brother-in-law gave the eulogy, and at some point during it, I heard wailing sounds coming out of myself. Incredible wails, like you see in the movies sometimes. I'd always thought those actors were over the top but realized then that they weren't. People really do wail. I was wailing. Chest heaving, animal sounds I'd never heard myself make.

Back in the math department at MIT, no one really said anything, even though I'd had to miss class for a few days to attend the funeral. But one person—Nancy Toscano, the administrative assistant who ran the department office—sent me a condolence card. I was, and still am, very touched by what she did, and surprised by the warmth and power of that simple gesture.

As for my math colleagues, I don't hold their silence against them. They didn't know what to say, just like I hadn't known what to say to Joff about Marshall's death.

In the face of all this, Joff's next letter came as a welcome balm.

Dear Steve, October 13, 1990

I just *have* to write you about this new academic year. I started with 7 students in a multivariable calculus course, am down to 6, and there will be no more attrition. What a wonderful collection of talented and motivated math scholars! Your name, even to the ones new to me as a teacher, has long since become a household word. (I doubt if one of them knows who Spiro Agnew is.) I must relate the anecdote which follows since it constitutes a very high water mark in classroom Founders 8:

Using liberally from the works of Woods that you sent me, and triggered by *Surely You're Joking, Mr. Feynman*, I showed them the differentiate under the integral "magic." The first efforts revealed $n! = \int_0^\infty x^n e^{-x} dx$. One lad came to class the next day armed with the Γ function (his father had supplied him with it). We saw it was a translation of the $n!$ (Euler?) function. In the dark recesses of my memory (I'm rambling: !@%X), I recalled you had sent me a graph of Γ which revealed min. $n!$. Somewhere I read Lord Kelvin remarked to a group that no one who didn't know $\int_{-\infty}^{\infty} e^{-x^2}\, dx = \sqrt{\pi}$ could consider himself a mathematician. Last year in multivariable calculus I was able to do the double integral, polar, which led to that result . . . sort of a summa cum laude climax to the senior year of mathematics.

I had started to tell a story. #!@%. This preface was beginning to have no end. Yesterday I took time in class to "bond" $\frac{\sin x}{x}$ to $e^{-\alpha x}$ and differentiate under the integral on $\int_0^\infty e^{-\alpha x} \frac{\sin x}{x}\, dx$ to obtain $\int_0^\infty \frac{\sin x}{x}\, dx = \frac{\pi}{2}$. One of my students was so amazed at the beauty of this that he sat there applauding for about half a minute! It was as though he had just seen Perlman conclude some impossible violin concerto which was being played for the first time.

Now don't include me in the performance. I was merely an enthusiastic carrier of this lovely piece of mathematics. I think this is the first time in my career that I have ever had a student react like he did. It does reassure me that the same potential for appreciation of excellence exists in math as much as in any other art that our young scholars take on.

#!@X **Nov.** 30, 1990

Things got very busy (fall exams, windsurfing, grades and comments, windsurfing), and this letter sunk to the bottom of all my paperwork. The multivariable calculus class continues to be a joy. Actually we haven't done much in 3-space yet (as the course name suggests); the fall has been delightfully spiced with Strogatz-Mirollo exercises and all the stuff we never got to in Calculus BC. A couple of these students recognized Rennie's name—they are on our math team and heard him lecture somewhere. They loved the Chinese restaurant place mat proof of $\Sigma \frac{\sin k}{k}$ you sent me. One of the students, David Choi, is *sure* he knows which restaurant it was. A former student (now math Ph.D. researcher John Hamilton for Kodak) sent me a letter suggesting a problem the NFL provided. He had noticed that some teams were intentionally taking a penalty before kicking a field goal in order to make their kicker feel better about his position relative to the goal posts—oops! I should mention that the ball is on a sideline hashmark. My guys tore into that one (maybe they aspire to become NFL team *mathematicians*) using vectors in 3-space to discover a surprising result. We reinforced that math by a "field" trip to the gridiron and let our visual sensations of the angle between goal posts do our lab work.

Speaking of legends, I showed them a proof that Jamie Williams did as a junior in my precal/cal class in early fall: that the altitudes of a Δ are concurrent. He used dot products in a neat little proof (which revealed he was well in command of the algebra of dot products). I think these guys would enjoy the downhill and uphill projectile shot, max range problems. I have toyed with doing a lab on an inclined table with a spring gun fixed to a quadrant, but then the reason I like teaching math is the absence of long lab set-ups and take-aparts. The gedanken experiment is a wonderful thing. I'm sure several guys could fire up their lap computers and do a non calc. solution of a given hill slope as they commuted to school one morning.

November has provided me with one of the best windsurfing months ever. There was a lot of Indian summer attended with fine winds and waves. I like to think that I am getting a hands-on feel for vectors when, a half-mile offshore in Long Island Sound, I overtake a 4-foot swell, hold for a calculus type instant at the top of the wave, and then drop down the face as my board speed accelerates to 20+ mph. One gets very tingly feelings of apparent wind in those situations! Yeah, you could say my batteries are charged for a while. (Do my students notice that silly inward grin in class as memories of an exhilarating vacation keep pushing to the surface, where the surface should be some serious stuff on surfaces of revolution?)

I'm afraid the twilight of my career approaches, but I am so grateful for the life that Loomis has given me. A regret is that I haven't been able to institute a math "hall of fame" here, but in fairness to the math greats that came before 1950 and those that I wasn't lucky enough to teach, I can only perpetuate the lore of the legends I knew.

Hope the holiday season brings you happiness and rest (do you need a break?).

Best regards,
Joff

A few days later came an unexpected diversion, a rare moment when it seemed like the whole country was arguing over a math puzzle. It all started when Marilyn vos Savant, author of the "Ask Marilyn" column for the Sunday *Parade* magazine (and listed in the Guinness Book of World Records Hall of Fame for "Highest IQ") answered a brain teaser posed by one of her millions of readers. The setup was based loosely on the old TV game show *Let's Make a Deal*, hosted by Monty Hall:

> Suppose you're on a game show, and you're given the choice of three doors. Behind one door is a car, behind the others, goats. You pick a door, say #1, and the host, who knows what's behind the doors, opens another door, say #3, which has a goat. He says to you, "Do you want to pick door number 2?" Is it to your advantage to switch your choice of doors?

Marilyn claimed that yes, you should switch because your odds of winning the car would then be 2/3, whereas if you stick with your original door, your odds would be only 1/3.

This answer was so counterintuitive that thousands of readers, including many professional mathematicians, sent her vehement letters. One wrote: "There is enough mathematical illiteracy in this country, and we don't need the world's highest IQ propagating more. Shame!" Another scolded: "As a professional mathematician, I'm very concerned with the general public's lack of mathematical skills. Please help by confessing your error and, in the future, being more careful." A third put it succinctly: "You are the goat!"

But in fact Marilyn was right. To see why, suppose you follow her strategy and switch when Monty gives you the chance. There are three possible ways the game could play out:

- You pick goat 1. Monty shows you goat 2. You switch. You win the car.
- You pick goat 2. Monty shows you goat 1. You switch. You win the car.
- You pick the car. Monty shows you a goat. You switch. You get the other goat.

So by switching, you win two out of three times, just as Marilyn said.

Another way to see intuitively why it's better to switch is to consider the limiting case of many doors, say a thousand of them. You pick door #1. Like every other door, it's a one-in-a-thousand long shot. Then Monty opens 998 doors with goats behind all of them, leaving only yours and one other, say door #723. "Now," he asks, "would you like to keep what's behind door #1, or switch to door #723?"

Can you see he's practically telling you where the car is? Switch to #723! Or do you really have so much faith in your guessing ability that you believe #1 is worth keeping?

Joff asked for my take on the controversy.

Hi Joff, January 7, 1991

I think Marilyn is right. After you pick a door and Monty (the game show host) reveals a goat, you will win 2/3 of the time if you switch to the remaining door. Her table of possibilities is a clear proof of this.

Some people are bugged at the apparent asymmetry of the two doors after Monty reveals a goat. But remember—*Monty does a nonrandom thing*: he always shows you a goat. So this is how the symmetry is broken. Still very confusing, however!

Two tidbits about Pascal's triangle:

(1) What is the "right" way to extend Pascal's triangle

1	level n = 0
1 1	n = 1
1 2 1	n = 2
1 3 3 1	n = 3
1 4 6 4 1	n = 4

when n is *negative*? Your students may find this amusing. (It's a bit open-ended.)

(2) Here's a neat way to find *Wallis's formula* for

$$\int_0^{2\pi} (\cos\theta)^{2m}\,d\theta,$$

where $m = \text{integer}$, $m \geq 1$. It uses Pascal's triangle and Euler's formula $\cos\theta = (e^{i\theta} + e^{-i\theta})/2$. We're faced with the monster integral

$$\int_0^{2\pi} \left(\frac{e^{i\theta} + e^{-i\theta}}{2}\right)^{2m} d\theta.$$

But wait! If we expand this with the binomial formula, all the integrals will vanish! (since we'll get a bunch of terms like

$$\int_0^{2\pi} e^{ik\theta}\,d\theta = \int_0^{2\pi} \cos k\theta\,d\theta + i\int_0^{2\pi} \sin k\theta\,d\theta = 0.$$

EXCEPT the $k = 0$ term (corresponding to the "middle term" in the binomial expansion) won't integrate to 0. When $k = 0$, $\int_0^{2\pi} e^{ik\theta}\,d\theta = \int_0^{2\pi} (1)\,d\theta = 2\pi$.

Now all we have to do is figure out the coefficient of the middle term. (Two cases we're used to are 2 for $(1+x)^2$ and 6 for $(1+x)^4$.)

1

1 1

1 (2) 1

1 3 3 1

1 4 (6) 4 1

Generally, we have

$$(1+x)^n = \sum_{k=0}^{n} \binom{n}{k} x^k,$$

where $\binom{n}{k}$ (read "n choose k") means $\frac{n!}{k!(n-k)!}$. The middle term occurs when we're halfway up the series to n (or halfway across a row of the triangle). For us, $n = 2m$, so we want the term $k = m$; i.e., the desired coefficient is

$$\frac{(2m)!}{m!m!} = \frac{(2m)!}{(m!)^2}$$

(since $(n-k)! = (2m-m)! = m!$). (Check: When $m = 1$, this gives 2, and when $m = 2$, it gives 6, as desired.)

OK, let her rip!

$$\int_0^{2\pi} \left(\frac{e^{i\theta} + e^{-i\theta}}{2}\right)^{2m} d\theta = \frac{1}{2^{2m}} \frac{(2m)!}{(m!)^2} \int_0^{2\pi} e^{i(0)\theta} d\theta$$

$$+ \text{ terms} \int e^{ik\theta}\, d\theta, \qquad k \neq 0$$

that integrate to 0.

The top integral on the right gives 2π. So we figure the answer is

$$\boxed{\frac{2\pi}{2^{2m}} \frac{(2m)!}{(m!)^2}} = \int_0^{2\pi} (\cos\theta)^{2m}\, d\theta.$$

You may want to play with the factorials to convince yourself that this really is Wallis's result

$$2\pi \cdot \frac{1}{2} \cdot \frac{3}{4} \cdot \frac{5}{6} \cdot \cdots \cdot \frac{2m-1}{2m}.$$

OK, Joff, that's all I have time for right now. Dinner's ready!

Adios amigo,
Steve

Infinity and Limits (1991)

In everyday language, infinity and limits sound like they contradict each other, but in calculus they are welded together as part of a single, overarching concept. In fact it is probably the most revolutionary concept in all of calculus, the one that distinguishes this branch of math from all that came before—algebra, geometry, and trigonometry.

The great achievement of calculus is the domestication of infinity. The concept of infinity itself had long been avoided and even feared. The Greeks thought it made no sense. During the Renaissance, the Church forbade anyone to write about infinity—and Giordano Bruno was burned at the stake for violating that edict—because only God could be truly infinite.

But the creators of calculus dared to confront infinity, because they had no other choice. Even before they could grasp it logically, they sensed that the infinite and its mirror image, the infinitesimal, were the keys to the unsolved mathematical problems of their era. Whether in the geometric problem of calculating tangents to curves, or in the physical problem of calculating the instantaneous velocity of a planet in orbit, the strategy was conceptually the same: take a complex motion or shape and treat it as an infinite number of infinitesimal parts. A curve becomes a collection of tiny straight lines; an elliptical orbit becomes a series of tiny excursions, each at constant speed and direction. The advantage is that the parts are more tractable than the whole. They change almost imperceptibly from one to the next, like the successive pages of a flip book.

The related concept of limit is inherently frustrating. It's exemplified by the old conundrum about moving half the distance to the wall, and then half of that distance, and so on. You keep getting closer but you never really get there.

The most fundamental notions of calculus are all phrased in terms of limits. Limits were introduced by the second wave of mathematicians studying calculus, to sidestep paradoxes that had confounded its creators (especially Isaac Newton) when they resorted to infinitesimals. Newton sometimes handled infinitesimals as if they were zero, and other times not, leaving him open to vicious (and surprisingly successful) charges of flip-flopping. His critics asserted that he was getting correct answers by making two logical errors that fortuitously canceled each other out, a method that struck them as untrustworthy. The eventual resolution came a century later. Newton's infinitesimals weren't actually zero; they were only approaching zero as a limit.

∫ ∫ ∫

In January and February of 1991, Joff and I corresponded frenetically, volleying letters back and forth faster than at any time before or since. The trigger was a cryptic statement about a limit that Joff had come across in Carl Boyer's classic book *A History of Mathematics*. Boyer was in the midst of deriving the equation

$$\frac{2}{\pi} = \frac{1\times1\times3\times3\times5\times5\times7\times7\times\cdots}{2\times2\times4\times4\times6\times6\times8\times8\times\cdots},$$

an "infinite product" formula that relates π to all the odd and even numbers. It's astonishing that π, which is about circles, should have anything to do with even and odd numbers, artefacts of arithmetic and seemingly remote from geometry. But that coincidence is not what bothered Joff. It was the way that Boyer had gotten to it.

En route to obtaining the infinite product, Boyer had casually made use of a certain limit as a number n approaches infinity. Specifically, he invoked the formula

$$\lim_{n \to \infty} \frac{\int_0^{\pi/2} \sin^n x \, dx}{\int_0^{\pi/2} \sin^{n+1} x \, dx} = 1,$$

which Joff found puzzling.

Joff and his class took a stab at proving Boyer's mysterious limit, without success. In his letter of February 7, he described their "free-swinging attack" on the problem and then off-handedly asked me about a rumor that I'd gotten engaged. When I wrote back on February 19, I showed him how to prove Boyer's limit by exploiting the fact that the graph of $\sin^n x$ becomes shaped like a spike as n approaches infinity. I said nothing either way about my rumored engagement. Elisabeth and I were clearly headed for the rocks, though the wedding was still on. Nor did I tell Joff about the unexpected and devastating death of my mother just a few months earlier.

First with his gentle question about my engagement, and later with the news that his youngest son was being treated for cancer, Joff seemed to be trying to change the unspoken rules between us, opening a new door in our relationship, or at least nudging it ajar. Whereas I, consciously or unconsciously, seemed to have been determined to stay in the well-ordered world of mathematics. Perhaps it was no accident that this round of our correspondence was devoted to limits and infinity. With mortality on our minds, maybe we were both seeking refuge in the one place where infinity becomes real.

Dear Steve, Monday, January 21, 1991

Your last letter sent me to our library to break out Boyer's *A History of Mathematics* so I could read about Wallis. I found that, according to the text, "after an evaluation of $\int_0^1 (x - x^2)^n\, dx$ for several positive integral values of n, Wallis arrived by incomplete induction at the conclusion that this integral is $(n!)^2/(2n+1)!$. Assuming that the formula holds for fractional values as well, Wallis concluded that $\int_0^1 \sqrt{x - x^2}\, dx = \left(\frac{1}{2}\right)^2/2!$" which led to $\frac{1}{2}! = \frac{\sqrt{\pi}}{2}$ (you led me to *other* factorials with the differentiate under the integral of Woods/Feynman methods earlier). Boyer then cited Wallis's

$$\frac{2}{\pi} = \frac{1 \cdot 1 \cdot 3 \cdot 3 \cdot 5 \cdot 5 \cdot \cdots}{2 \cdot 2 \cdot 4 \cdot 4 \cdot 6 \cdot 6 \cdot \cdots}$$

as one of his best known results and said it could be obtained from the modern theorem

$$\lim_{n \to \infty} \frac{\int_0^{\pi/2} \sin^n x\, dx}{\int_0^{\pi/2} \sin^{n+1} x\, dx} = 1$$

and

$$\int_0^{\pi/2} \sin^m x\, dx = \frac{(m-1)!!}{m!!} \quad \text{for } m \text{ an odd integer and}$$

$$\int_0^{\pi/2} \sin^m x\, dx = \frac{(m-1)!!}{m!!} \cdot \frac{\pi}{2} \quad \text{for } m \text{ an even integer.}$$

I feel comfortable with the latter two as a result of your showing me $\int_0^{\pi/2} \cos^{2m} x\,dx$, but that first limit has me puzzled. I saw !! defined for the first time:

$$m!! = m\,(m-2)\,(m-4)\ldots$$

ending in 2 or 1. Where have I been?

I worked out $\int_0^1 (x - x^2)^n\,dx$ for $n = 1, 2, 3,$ and 4, but I can't say the sequence of numbers screamed $\frac{(n!)^2}{(2n+1)!}$. At least my results fit the fractional formula, and my bag of tricks for fitting a formula to a table of values may be limited to finite differences for polynomials.

This Persian Gulf business is depressing. I am reminded of Robert Ardrey's inference that mankind has *always* killed mankind when he discovered a crushed skull in an Olduvai Gorge dig. Young Leakey took him to task about that sort of nonscientific speculation, but our aggression within our species does seem rather singular in the natural world.

Why would anyone wish to fight if he could go windsurfing in Long Island Sound on an almost balmy Sunday 20 Jan '91 in sparkling blue waters and 12 mph breezes which knew not to exceed the windchill tolerance of this sailor!

Best,
Joff

Dear Steve, Thurs. Eve. February 7, 1991

We made an attempt at

$$\lim_{n\to\infty} \frac{\int_0^{\pi/2} \sin^n x\,dx}{\int_0^{\pi/2} \sin^{n+1} x\,dx}.$$

One of my lads was very quick to see that, with the exception of $x = \frac{\pi}{2}$, escalating powers of $\sin x \to 0$. I was ready for this and produced the graph of $\sin^{10} x$, $\sin^{100} x$ from my calculator. Another student observed that this limit seemed to take on the L'Hôpital look of $\frac{0}{0}$, but things stalled there. I got the idea to push on this intuition and suggested a free-swinging attack of giving the limit a "layered look":

$$\left[\lim_{\substack{t \to \pi/2 \\ n \to \infty}} \frac{\int_0^t \sin^n x \, dx}{\int_0^t \sin^{n+1} x \, dx}\right], \qquad \text{where } 0 < t < \frac{\pi}{2},$$
$$= \lim_{\substack{t \to \pi/2 \\ n \to \infty}} \frac{\sin^n t}{\sin^{n+1} t} = 1.$$

Never having done anything like this, I wonder if it is mathematical? Getting a result we were pushing for doesn't give me any encouragement. Once again I find myself operating in a groping way, wondering what a professional mathematician would say if confronted by these attempts !X%@.

We feel rather comfortable with Wallis's $\int_0^{\pi/2} \sin^m x \, dx$ (m an even natural number) $= \frac{\pi}{2} \frac{(2m-1)!!}{(2m)!!}$ since your method for $\int_0^{2\pi} \cos^{2m} x \, dx$ prompted us to depart with $(\frac{e^{ix} - e^{-ix}}{2i})^m$ and see $\int$'s get zapped. In the recesses of my mind is the letter you sent me about Fourier square waves and sawtooth waves in which orthogonality zapped integrals. At the moment I have mislaid your method of finding the Fourier coefficients. If I remember, your method was presented so I could understand it. Looking in my Agnew calculus text, his explanation was beyond me. There was a déja vu sensation that I'd done things before (and forgotten them).

The $\int_0^{\pi/2} \sin^m x \, dx$ (m odd) $= \frac{(2m-1)!!}{(2m)!!}$ has stumped us. We can see the pattern developing by looking at $m = 1, 3, 5, 7$, but in this formula there are a lot of nonvanishing integrals to be summed, and I don't have enough time to beat on the sum, which I think needs help from summing $\binom{n}{r}$-type critters.

In any case, amidst the usual routine of doing partial derivatives and soon the (perplexing) chain rule for multivariable work, we have enjoyed tromping through grounds where we have no experience, if only for a break in the routine.

Today I did $\int_0^{\pi/2} \sin x\, dx$ the *long* way, from $\int_0^{\pi/2} \frac{e^{ix} - e^{-ix}}{2i}\, dx$ and integrating the terms like $\int e^{ix}\, dx = \frac{1}{i} \int e^{ix} i\, dx = -ie^{ix}$, by $\int e^u\, du$ as if these were ordinary real functions. We were pleased that $\int_0^{\pi/2} \sin x\, dx = 1$, even the long way. I have had a complex analysis book by my bedside, but one doesn't get very far when beginning at 11:30 p.m. with one's eyes already drooping. I regret never studying the subject in college, now, but I never saw calculus until college either.

The West Hartford day student rumor mill says you are engaged. Evidently, my student Jordy Oland knows the Karp family well, and there was a leak.

We are all happy for you, Steve (if this rumor is true).

I spent Sunday at sea in Long Island Sound on my sailboard, getting a firsthand feeling for the bliss that comes from the vectors of wind, wave, and gravity combining into a sailing serendipity. And I didn't have to share this with anyone. Except for a tanker on the distant horizon, I was the sole possessor of a kingdom which by summertime would be crowded with power boats, jet skis, and assorted other craft.

Hope all goes well with you, Steve; do you have a spring break? If you're in the vicinity of the homestead, we'd love to have you visit.

Best,
Joff

Hi Joff, February 19, 1991

Another busy day, so I'll be brief. Here are some thoughts about

$$\lim_{n\to\infty} \frac{\int_0^{\pi/2} \sin^n x \, dx}{\int_0^{\pi/2} \sin^{n+1} x \, dx}.$$

Let's look at a related integral:

$$\int_{-\pi/2}^{\pi/2} \cos^n x \, dx.$$

The integrand $\cos^n x$ looks like this for large n (I'm using cosine because $x = 0$ is nicer than $x = \frac{\pi}{2}$ for what I'm about to show):

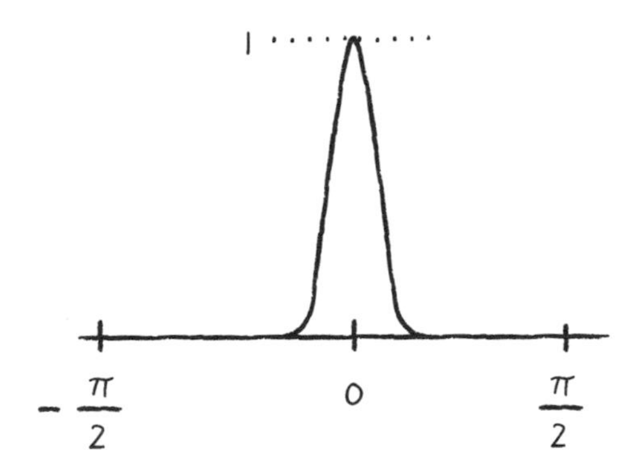

The only part of the function that contributes significantly to the integral is the spike near $x = 0$. So if we can successfully approximate that spike by something simpler, we'll be in good shape.

There's a standard trick for dealing with such "sharply peaked" functions, called Laplace's method. Near $x = 0$, we have

$$\cos x = 1 - \frac{x^2}{2} + O(x^4).$$

But

$$1 - \frac{x^2}{2} + O(x^4) = e^{-x^2/2}$$

since the error is $O(x^4)$; in other words, $e^{-x^2/2}$ and $\cos x$ are practically equal for small x.

Why is this helpful? Because

$$(\cos x)^n \approx (e^{-x^2/2})^n = e^{-nx^2/2}$$

is a very good approximation when n is large! In other words, $e^{-x^2 n/2}$ is an excellent approximation to the spike; for x outside a window of width $|x| \lesssim O(\frac{1}{\sqrt{n}})$, the function $e^{-x^2 n/2}$ is tiny anyway, and so the error we're making is negligible as $n \to \infty$. To put it more simply, *have your students graph* $(\cos x)^n$ *and* $e^{-nx^2/2}$ *for* n *large*. You'll see the two functions are very close! (I hope!) This should work great on the interval $\left[-\frac{\pi}{2}, \frac{\pi}{2}\right]$. Of course when $x = 2\pi$, the approximation will be lousy, but we're not going out that far!

The point is, we can replace $\int_{-\pi/2}^{\pi/2} \cos^n x\, dx$ with $\int_{-\pi/2}^{\pi/2} e^{-nx^2/2}\, dx$ and make a negligible error for n large.

But what's the use of this, you ask? We can't do the integral $\int_{-\pi/2}^{\pi/2} e^{-nx^2/2}\, dx$ either. *But* now the brilliant move—extend the range of integration to $\int_{-\infty}^{\infty} e^{-nx^2/2}\, dx$. That includes more of the function $e^{-nx^2/2}$, but it's so extremely tiny that it doesn't matter! Only the region near the spike is important.

Now we know how to find $\int_{-\infty}^{\infty} e^{-nx^2/2}\,dx$ exactly using the polar coordinate trick. So I'll trust you can verify $\int_{-\infty}^{\infty} e^{-nx^2/2}\,dx = \sqrt{2\pi}\frac{1}{\sqrt{n}}$.

So here's our reasoning:

$$\int_{-\pi/2}^{\pi/2} (\cos x)^n\,dx \sim \int_{-\pi/2}^{\pi/2} e^{-nx^2/2}\,dx \sim \int_{-\infty}^{\infty} e^{-nx^2/2}\,dx$$
$$\approx \sqrt{2\pi}\frac{1}{\sqrt{n}}$$

since only the spike matters. Taking half of this produces the following picture:

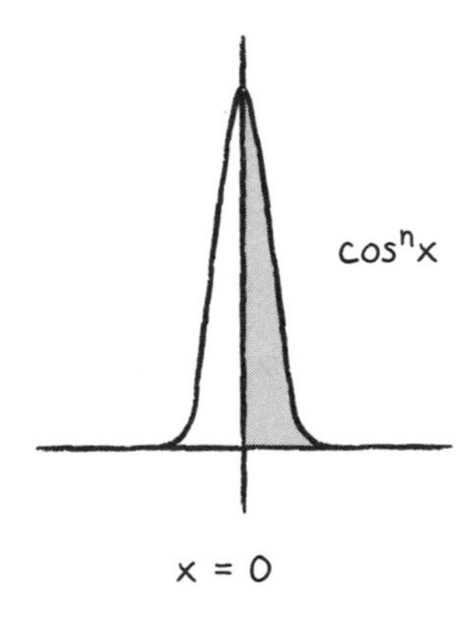

$$\int_0^{\pi/2} \cos^n x\,dx \approx \frac{1}{2}\sqrt{2\pi}\frac{1}{\sqrt{n}}$$

$$\|$$

$$\int_0^{\pi/2} \sin^n x\,dx \text{ by geometry.}$$

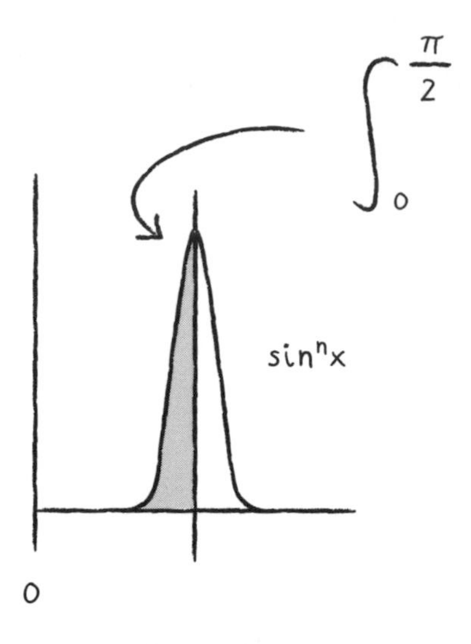

So I claim your integral

$$\boxed{\int_0^{\pi/2} (\sin x)^n \, dx \approx \sqrt{\frac{\pi}{2}} \frac{1}{\sqrt{n}}}$$

for large n. Let's see if this is any good, using our computer.

	$\int_0^{\pi/2} \sin^n x \, dx$	$\sqrt{\frac{\pi}{2}} \frac{1}{\sqrt{n}}$
$n = 1$	1	1.25
2	.78	.89
3	.67	.72
4	.59	.63
		(not bad ↑ for small n)
⋮		
99	.1256	.1260

The numbers .1256 and .1260 in the last row agree to 0.3%. So agreement looks good. *Make a graph of this table.*

Anyway, now it's clear:

$$\frac{\int \sin^n x\,dx}{\int \sin^{n+1} x\,dx} \approx \sqrt{\frac{n+1}{n}} \to 1 \qquad \text{as } n \to \infty.$$

QED

The simpler point is that

$$\begin{aligned} \sin^{n+1} x &= \sin^n x \sin x \\ &\approx \sin^n x \quad \textit{in the region of the spike} \\ &\qquad\qquad \text{since } \sin x \approx 1 \text{ there.} \end{aligned}$$

So it makes sense that the ratio of the integrals is 1.

Cheerio,
Steve

Dear Steve, Monday February 25, 1991

You are amazing! Your bundle of mathematics arrived today with the exciting notions of spikes and the $(\cos x)^n \sim (e^{-x^2/2})^n$ revelation, as well as (and here is where I am really embarrassed) the 14 Mar '89 volume which led to the Fourier methods which I have misplaced.

My multivariable students will enjoy new flights on mathematical wings thanks to your launching them from the nest (again!). I'll enclose yet another *Parade* magazine "Ask Marilyn" column in case you haven't seen it. This counterintuitive game show problem does not die easily. I wonder if you know Seth Kalson, who wrote what may be the final response letter to Marilyn.

The Joff family has been sweating out a problem since Christmas. Our youngest son Jeff (24) has been operated on for cancer and has just completed a 3rd stage of chemotherapy. We finally got some encouraging news today that his blood "markers" are back to normal. Jeff comes home today, and we are hoping he'll not have to undergo a 4th chemotherapy treatment. The crocuses just came out today, so maybe good things are on the way.

I am up to my ears with exams, but winter term ends Friday, so a chance to recharge my batteries and fuss with the forgotten Fourier coefficients method looms brightly. I did try to zap all but the appropriate coefficient by integration, but something wasn't working. I am deliberately holding off reading your explanation, hoping for some last minute neuron stimulation (very possibly the ones entrusted to this secret have died! X%@!).

Best regards,
Joff

Chaos (1992-95)

A dynamical system is anything that changes from moment to moment according to definite rules. For instance, a pendulum swinging back and forth obeys Newton's laws of motion and the law of gravity. Using those rules and the techniques of calculus, physicists can add up all the moment-to-moment changes to deduce where the pendulum will be at any time and how fast it'll be moving. The same is true for a planet orbiting the sun or a rocket flying to the moon. Other scientists have extended the reach of dynamical systems to describe the flow of traffic on the Internet, the fluctuations of wildlife populations, and the beating of the human heart.

All along, the implicit assumption was that the moment-to-moment predictability of dynamical systems meant they should also be predictable forever, at least in principle. Now we know that's not true.

Some dynamical systems can be chaotic.

Here, chaos doesn't mean utter confusion. It means that even though the system evolves according to definite rules, you still can't predict what it'll do in the long run. Why not? Because chaotic systems are exceedingly sensitive. Any little uncontrolled disturbance—the proverbial flap of a butterfly's wings—gets amplified so rapidly that the system behaves completely differently from how it would have otherwise. The errors in any attempted forecast start to snowball exponentially fast, rendering prediction meaningless. There's no way around this. It's not a matter of needing better instruments or being more careful or waiting for better mathematical methods to come along. Chaos is an unavoidable part of reality.

It's also a slap in the face. Scientists, like everyone else, had always known that complicated things like human relationships or wars or history could be unpredictable. But at least we had our pendulums for comfort. Now chaos was taking those away too.

$\int \int \int$

The applied math group at MIT had hired me to be their chaos expert. I soon began writing a textbook on the subject, which provided a haven from my real life. Elisabeth and I had gotten married in June 1991, willfully ignoring the warning signs. Our relationship improved for a while, eventually stabilizing into a placid friendship. Not intolerable, but a paler version of marriage than either of us had hoped for. The analytical discussions of our counseling sessions were a highlight of the week for me, whereas Elisabeth found them tedious. She was ready to move on. We filled out a few legal forms she had found online, stood before a judge, and that was that.

Between the activity at work and the unraveling at home, I found little time to correspond with Joff. He wrote to me seven times between the fall of 1991 and the spring of 1995. I wrote to him twice. Neither in those two letters, nor when I look through any of my notes, is there any record of my ever inquiring about his son's cancer. My self-absorption staggers me when I reflect on it now.

A decade earlier I'd written to him about my dreams of a linear life, about how I'd hoped to get married someday, become a professor, and live happily ever after. The first part had already failed. At least there was work. I loved MIT and hoped to make my career there. The problem was the matter of tenure. Nobody had gotten tenure in applied math at MIT for a very long time. The last one who had made it was teased by his older colleagues. "You were lucky," they'd say, "you didn't deserve it." Young professors would arrive, fantasize that maybe they'd be the one, and then be dismissed a few years later. We all knew what was in store for us, but it was still demoralizing.

Nick Trefethen, a brilliant numerical analyst a few years older than me, was led to believe he had a chance. See, it *is* possible, we all thought. Feeling optimistic, he went so far as to buy a house near his parents in Lexington and moved his wife and kids there. Then the word came down that no, actually, he would not be given tenure after all. (Today he's a professor at Oxford and a Fellow of the Royal Society.)

When my turn came, the department put me up for promotion to associate professor without tenure, the next step up the ladder. A few weeks later I received a letter on plain white paper, no letterhead, in a plain white envelope marked "Confidential." There were two paragraphs. The first informed me I was being promoted. The second read as follows:

> At present your research does not constitute adequately deep mathematics and, as a result, your present prospects for tenure at MIT are low. Of course, it is possible that future outstanding research work could change this assessment. However, on the basis of your work to-date it is necessary to advise you that the tenure outlook at this time is not positive.

Faced with these odds, I applied for jobs at other top institutions, hoping that a competing offer might change MIT's mind. Cornell, arguably the best place in the country for chaos theory, offered me tenure two years early. That got MIT's attention. They conducted another assessment, informally this time. While advising me to consider Cornell's offer seriously, they also told me they hoped I'd stay and encouraged me to try to "hit a home run" (the phrase they always used) so that tenure might go through in two years time.

I agonized over this. What I feared most was that, if I did move, would I ever find someone to marry? The social prospects for young professors in Ithaca looked bleak.

I called my brother Ian. "Steven, this is not a tough decision," he said. "Cornell is a great school and they're making you a terrific offer. And don't worry, there are plenty of women in Ithaca. You'll be fine."

When I finally resumed my correspondence with Joff in the spring of 1995, I'd been at Cornell for a year already. I mentioned nothing about my divorce or what had happened at MIT. Instead I dove straight into a calculation about an infinite product. But before sealing the envelope, it occurred to me that Joff would wonder why I was now writing to him from Cornell. So I added a sanitized explanation as an afterthought.

Dear Joff, March 21, 1995

Thought it was about time (actually long overdue) to resume our correspondence. Here's an infinite product formula you might like. It is a nice application of the trig identity

$$\sin 2x = 2\cos x \sin x$$

and the limit formula $\frac{\sin x}{x} \to 1$ as $x \to 0$.

Suppose we start with the "half-angle formula"

$$\boxed{\sin x = 2\sin\frac{x}{2}\cos\frac{x}{2}.}$$

(To get this, replace x by $\frac{x}{2}$ in the identity above.) We suddenly get greedy and decide to apply this formula to the factor $\sin\frac{x}{2}$ on the right hand side. Thus, the same reasoning yields

$$\boxed{\sin\frac{x}{2} = 2\sin\frac{x}{4}\cos\frac{x}{4}.}$$

Hence, after substitution,

$$\begin{aligned}\sin x &= 2\sin\frac{x}{2}\cos\frac{x}{2}\\ &= 2\left(2\sin\frac{x}{4}\cos\frac{x}{4}\right)\cos\frac{x}{2} = \left(4\sin\frac{x}{4}\right)\left(\cos\frac{x}{2}\cos\frac{x}{4}\right).\end{aligned}$$

Do this again, now using $\sin\frac{x}{4} = 2\sin\frac{x}{8}\cos\frac{x}{8}$ to get

$$\boxed{\sin x = 8\sin\frac{x}{8}\left(\cos\frac{x}{2}\cos\frac{x}{4}\cos\frac{x}{8}\right).}$$

We are obsessed! Keep using

$$\sin\left[\frac{x}{2^{n-1}}\right] = 2\sin\frac{x}{2^n}\cos\frac{x}{2^n}.$$

You see the pattern. If we keep doing this trick, we generate the formula

$$\boxed{\sin x = 2^n \sin\left(\frac{x}{2^n}\right)\prod_{k=1}^{n}\cos\left(\frac{x}{2^k}\right).}$$

(Recall that $\prod$ is the product symbol, like $\sum$ is a summation.) What happens as $n \to \infty$? We see that the argument $\frac{x}{2^n}$ of the $\sin(\frac{x}{2^n})$ goes to 0 for fixed x. This suggests that we want to divide both sides of the boxed equation above by x, to get

$$\frac{\sin x}{x} = \underbrace{\frac{2^n}{x}\sin\left(\frac{x}{2^n}\right)}\prod_{k=1}^{n}\cos\left(\frac{x}{2^k}\right).$$

The term marked by the underbrace is of the form $\frac{\sin u}{u}$, where $u = \frac{x}{2^n}$. As $n \to \infty$ we have $u \to 0$ and $\frac{\sin u}{u} \to 1$. Hence

$$\boxed{\frac{\sin x}{x} = \prod_{k=1}^{\infty}\cos\left(\frac{x}{2^k}\right)}$$

(and our derivation shows this holds for all x). This is the promised infinite product!

For further fun, you should look at $x = \frac{\pi}{2}$:

$$\frac{\sin\frac{\pi}{2}}{\frac{\pi}{2}} = \boxed{\frac{2}{\pi} = \prod_{k=1}^{\infty}\cos\left(\frac{\pi}{2^{k+1}}\right).}$$

Find formulas for these cosines:

Note $\cos\frac{\pi}{4} = \frac{\sqrt{2}}{2}$. Now write $\cos\frac{\pi}{8}$ using

$$2\cos^2\frac{\pi}{8} - 1 = \cos\frac{\pi}{4}; \quad \text{i.e.,} \quad \cos\frac{\pi}{8} = \sqrt{\frac{1+\cos\frac{\pi}{4}}{2}}$$
$$= \sqrt{\frac{1+\frac{\sqrt{2}}{2}}{2}} = \frac{\sqrt{2+\sqrt{2}}}{2}.$$

Observe the pattern:

$$\cos\frac{\pi}{4} = \frac{\sqrt{2}}{2}$$
$$\cos\frac{\pi}{8} = \frac{\sqrt{2+\sqrt{2}}}{2}.$$

You can show that $\cos\frac{\pi}{16} = \sqrt{2+\sqrt{2+\sqrt{2}}}/2$ using $\cos\frac{\pi}{16} = \sqrt{\frac{1+\cos\frac{\pi}{8}}{2}}$, etc. The result is a formula for π (actually $\frac{1}{\pi}$) in terms *solely of 2s and square roots!* The final formula is something like this:

$$\frac{2}{\pi} = \cos\frac{\pi}{4}\cos\frac{\pi}{8}\cos\frac{\pi}{16}\cdots$$
$$= \frac{\sqrt{2}}{2}\cdot\frac{\sqrt{2+\sqrt{2}}}{2}\cdot\frac{\sqrt{2+\sqrt{2+\sqrt{2}}}}{2}\cdots.$$

Enough for now.

Hope all's well. Life at Cornell is good. (Wait! Have we talked about how I came to be here? Cornell offered tenure 2 years early and made an irresistible offer!) I'm gradually making the transition from 12 years of a more hectic life in Boston/Cambridge. Ithaca has its charms. And I'm enjoying my colleagues and students here.

Hope you're well. What's up?

Warmest regards,
Steve
TAM Dept.
Kimball Hall
Cornell University
Ithaca, NY 14853
607-255-5999 (w)
607-257-5911 (h)

TAM stands for Theoretical and Applied Mechanics—essentially Cornell's version of Applied Math.

Celebration (1996-99)

On April 22, 1999, I was invited to pay tribute to Joff at a special event. Called "An Evening in Celebration of Education," it was a black-tie ceremony held in the Cathedral Church of Saint John the Divine in New York City. More than 100 alumni, faculty, parents, and friends gathered in that breathtaking space to celebrate teaching and to honor Joff in particular.

It must have been a bittersweet moment for him. Each spring he'd confront the decision of whether to sign on for another year, and until now the answer had always been yes. As he once wrote to me: "There is still that warming glow of adrenalin every time I cross the threshold of my classroom, and who wouldn't want that?"

But now, after 49 years of service—a school record—he was ready to stop.

I was relieved that Carole was able to accompany me to the event. We'd met in 1997, fallen in love on the spot, and gotten married in 1998. Now here we were, sitting next to Joff and his wife Sue at the head table, enjoying a great dinner and many laughs. Joff was understandably buoyant, maybe a little embarrassed, but not so embarrassed that he couldn't sing a few corny show tunes when it was his turn at the podium.

My mood was less relaxed. Memories of an earlier event flooded back, of that time in the school dining hall when I was overcome with emotion and couldn't finish my speech. On that occasion I hadn't prepared properly. This time, I had a new strategy: I solicited anecdotes from many of Joff's former students and planned to read the speech, not wing it. The hope was that whatever I might lose in spontaneity I'd gain in composure.

It went well. "My cup runneth over. . . . This will always be a high point in my career," Joff wrote a few days later.

His thank-you note was overflowing, not just with expressions of gratitude but also with stories about the weight throwers he was coaching on the track team, the jazz gigs he was playing, and the like. On the other hand, and though he tried to make light of it, he confessed to feeling apprehensive about his coming descent into retirement:

The math department surprised me the other day by putting my name on the senior math prize. I am getting nervous about all this great fortune. At Wesleyan I read a ballad by Schiller called *The Ring of Polykrates*. King Polykrates was daily greeted by great news . . . battles won, new land added to his kingdom . . . his wiseman took note of all this good fortune and suggested he get rid of his most treasured possession, so the king threw his ring into the ocean. The wise man looked relieved. The next day, dinner (fish) was served to the king and his ring fell out as the fish was sliced open for serving. The wiseman blanched and took leave, never to return.

I feel as though the gods are setting me up for some less than attractive event, but I'm not so superstitious as to throw my most treasured possession in the ocean. Besides, Sue doesn't like swimming in Long Island Sound when the water temperature is barely 50°F.

The Path of Quickest Descent

(2000-2003)

Joff's opening words when he resumed our correspondence were "Withdrawal from math teaching hasn't been easy."

Over the next several years he inundated me with letters, one after another without waiting for a reply. He'd devise his own math problems and show me his solutions, just like I had when I was his student. He'd write about nature and his vacation trips and tell me about his new friends, people I'd never met or heard of.

Mostly I was silent and didn't write back, too busy now with our first daughter. Then another. Never enough sleep. Helping Carole. Writing a book. Chores at school. I couldn't keep up.

He could sense what was happening. He started punctuating his letters with apologies. After deriving a formula for how much spherical area a bird can view on the surface of the earth as a function of its altitude, he wrote:

Well, Steve, I hope I haven't burdened you with this stuff. I treasured our correspondences of the past so much that I hate to let them go. And I didn't want you to think that all I'm doing these days is playing around in Long Island Sound in my kayak.

In another, after trying to invent a tetrahedral fractal surface that generalizes the Koch snowflake curve, he wrote:

I know you'd never have any trouble with this. I show you my reasoning only for your scrutiny and as a check on my calculations. Do you sense I have a 72-year-old's concern about losing "it" to senility and Alzheimer's?

The guilt was hard to bear. I took half-measures, which made me feel even worse:

Wish I could write you a proper letter but with Leah and my limited sleep and too many chores at school, this is going to be a poor excuse for a reply. I've enclosed a copy of my lecture notes for a class I taught recently about Bayes's formula. . . .

But the deluge continued. Joff wanted to let me off the hook, and yet he couldn't stop writing. "Enough already!" he wrote in closing another letter. "*In no way* feel a need to respond. Just know that a moment of writing you is a very special moment for me."

∫ ∫ ∫

The next letter shows how hard it was for Joff to let go of his earlier life. He recounts how he buttonholed a good-natured waitress and taught her about the path of quickest descent, a classic problem in the calculus of variations.

The problem goes like this. Consider two points *A* and *B* in the same vertical plane and suppose you want to connect them with the best possible chute. Your goal is to design the chute so that a frictionless particle will slide down from *A* to *B* in the shortest possible time.

You might think you should connect them with a straight line:

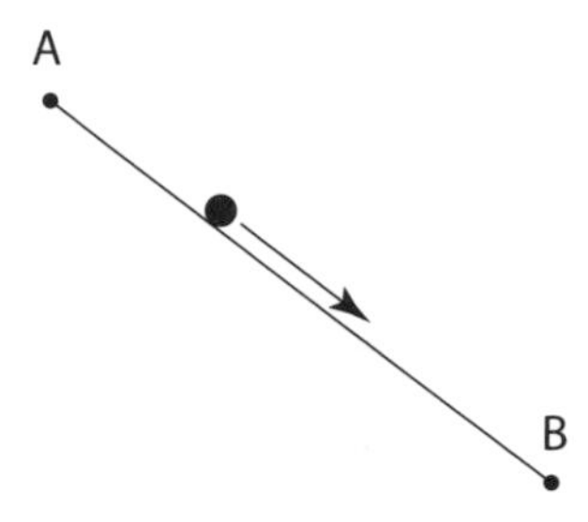

since that's the shortest path. On the other hand, maybe it would be better if the chute went down more steeply at first, to help the particle pick up speed, with the hope that the extra velocity would more than make up for the greater distance traveled.

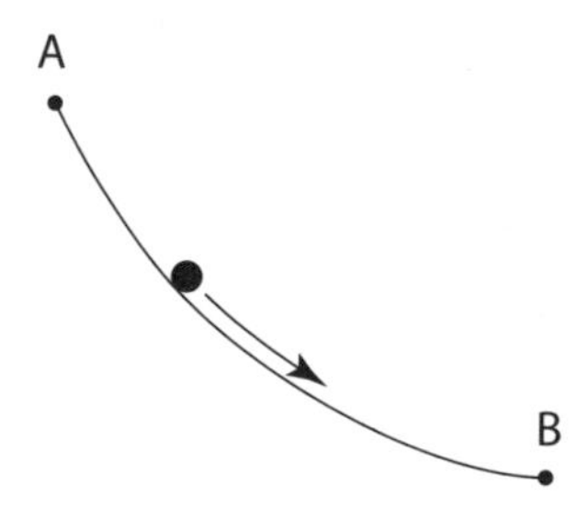

In that case, what's the right shape of the curved path?

The answer turns out to be the arc of an inverted cycloid. You've seen a cycloid if you've ever been mesmerized by the glowing trace of a little light bulb on the wheel of a bicycle, bobbing up and down as the cyclist rides along a darkened road at night.

The cycloid has another intriguing property. Besides being a "brachistochrone" (a path of shortest time), it's also a "tautochrone"—a path of *equal* time. What this means is that if you place a particle on an inverted cycloid and release it from rest, it always takes the same amount of time to reach the bottom, no matter where on the cycloid you release it. If you start it up higher, it has to slide farther, but it also moves faster—just the right amount faster to compensate for the extra distance. Picture it as a race between two particles starting at different points on the chute. The outcome is always the same: a tie. The two particles cross the finish line at the bottom in a dead heat.

Dear Steve, Monday, Sunrise without the sun, 7 Jan. '02

Sue and I went to check our house on West Hill Pond yesterday. Despite it being January there was no snow and the pond was about 80% ice-free . . . just a skim of ice on the north end. Sue elected not to go canoeing! It would have been a first for us, and we would have had the lake all to ourselves.

After dropping off some firewood and kindling we went to Manchester to celebrate daughter-in-law Jen's birthday by having dinner at the MACARONI GRILL. This is a favorite restaurant of mine since they have paper table cloths and the waitress signs in with a crayon (they practice doing their signature upside down to dazzle the patrons).

During the evening I preempted a crayon, and fortified by a glass of chianti, decided to derive the integral which, when solved, demonstrates the tautochrone. Hey! It's thanks to you that I've been seeing if I remembered the calculus from what has been over 3 years' retirement now. The waitress was fascinated at the math which was filling my place setting. I gave her a brief history of the path of quickest descent. She confessed she would have selected the straight-line path. I asked if she was a downhill skier. Getting an affirmative, I asked her to imagine she was on a mountain and noticed an attractive guy just leaving the lift and starting downhill along a linear slope. Were she to choose a descent along an inverted cycloid, she'd be at the bottom of the slope before him . . . and possibly share a chair with him on the ride back up.

In the rude light of day, I see my crayon creation has a few flaws. Somewhere I lost the constant $\sqrt{a/g}$, which should be parked outside the integral. Oh, the waitress smiled vicariously when I described the thrill of seeing θ_0 drop out of the integral when the limits $\int_{\theta_0}^{\pi}$ are chosen.

For you, Carole, and Leah,
Best wishes for '02,
Joff

Joff's creative outlets drew inspiration, as always, from the world around him. He started decorating his letters with color sketches of animals, plants, and people. He concocted math problems about birds he'd seen or boat houses he'd built or, in the next letter, how to mark volume levels on a cylindrical tank, prompted by a caller's question on the radio show *Car Talk*.

Dear Steve, 12/14/02

A friend told me he'd heard that radio show with the 2 car-repair brothers (the one that gave us the "label 15 wires crossing under a highway" challenge). Someone had called in wondering how he should mark volume levels on his cylindrical tank. The reply was "That's a calculus question."

I set about trying to find a noncalculus solution and came up with a precal way, with a simple (nongraphing, noncomputing) calculator assist. This adventure, back to topics taught in yesteryear, was quite enjoyable. For your amusement I'll show you what I came up with:

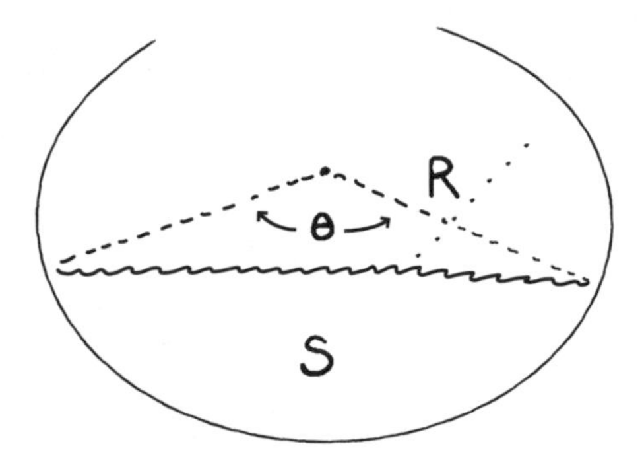

Region S has area $\frac{R^2}{2}(\theta - \sin\theta)$. . . . I loved this neat formula which has a simple elegance thanks to radian measure.

When the tank is $\frac{1}{4}$ full, $S = \frac{\pi R^2}{4}$, so $\theta - \sin\theta = \frac{\pi}{2}$, or

$$\theta - \frac{\pi}{2} = \sin\theta,$$
$$\theta - \frac{\pi}{2} = \cos\left[\frac{\pi}{2} - \theta\right] = \cos\left[\theta - \frac{\pi}{2}\right].$$

Let $x = \theta - \frac{\pi}{2}$, solve $x = \cos x$.

For this I grabbed my calculator and used the fixed-point method to approximate the solution.

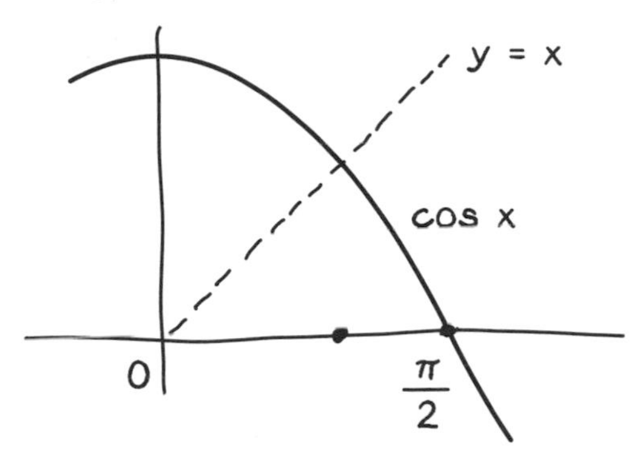

I guessed 0.7 would be a good seed number, entered 0.7, and then repeatedly pressed COS until my calculator alternately converged on 0.7390851. So $\theta = x + \frac{\pi}{2} \approx 2.3098814$ (sig. #'s !).

Home stretch:

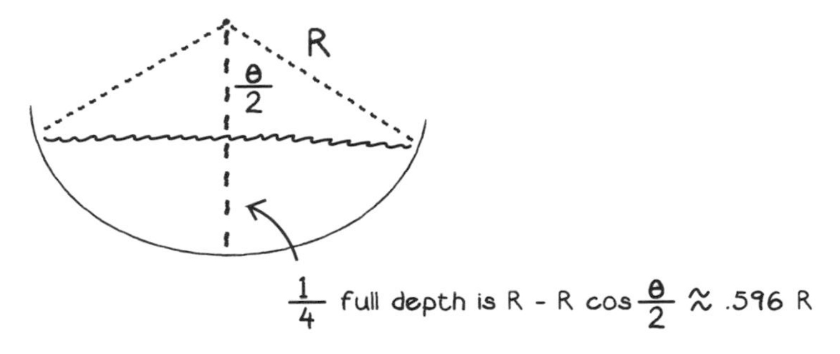

I hope the tank owner will accept my stick markings:

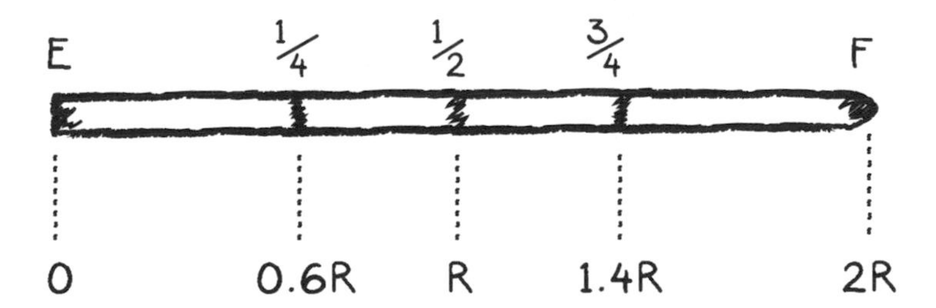

Aftermath: In my haste to sketch the guy on page one I'm afraid I created a small error. If this guy wasn't sharp enough to figure out the markings on the stick, he can be forgiven for wearing his left glove on his right hand. Don't ask how he is grasping the stick. Maybe I need an assist from Escher.

I've been playing Santa's elf in my workshop, building presents. At the moment I'm on the task of making a prefab BAT HOUSE kit for my son, Rex, and his son, Jesse, to assemble. The parts will be color-coded and all nail holes predrilled to ease the assembly chores for 2nd grader Jesse.

I just finished painting a mailbox for Ellie, a single gal who lives down the street. She wanted to get a larger mailbox. Would I paint it with some choice of local flora or fauna? I chose a hooded merganser, since we'd had a pair of them in our marsh pond for a week. Ellie wants *both* the drake and hen, one on each side. I obliged, and after several days of painting, and an hour of installation (the old mounting was quite resistant to my attempts to remove the rusted screws), I looked with pride on my efforts. Best of all, my acrylic paints survived a heavy rainstorm that first night.

Wreath making occupies much of my time. I built a LARGE (5-foot) wreath for the Loomis homestead again. This now has become a tradition which I started in the latter years of our living there. The wreath covers the double-door front entrance, but the residents always use the side door for coming and going.

Winter kayaking gives me a chance to frequent the Sound, the tidal rivers, and the marsh creeks in a serene solitude. Gone are the fair-weather boaters; it's just me, the ducks, herons, hawks, gulls, and swans . . . and gentle snowflakes descending from Payne's gray clouds.

Sue and I send our best wishes to you, Carole, Leah, and Joanna for the holidays. May your year 2003 contain 365 great days!

Joff

Bifurcation (2004)

Change can occur in varying degrees of violence and unpredictability. Accordingly, there are different kinds of mathematics for each.

At the mildest extreme lie the orderly changes of a system obeying differential equations, gliding along according to laws of motion. Think of the planets orbiting the sun or the burbling of a brook. Here calculus is in its element. It is designed for systems that evolve according to definite rules.

At the opposite extreme lies wild, irrational change. The kind that comes from senseless shocks to a system, external events bearing no logical relation to the system itself. The asteroid that hit the earth and wiped out the dinosaurs would be an example. The logic of the earth's ecosystem instantly became irrelevant. Calculus can make no sense of such situations. No mathematics can. Probability theory tries its best, by enumerating the statistics of chance events, but its predictive power in any specific situation is limited.

Between these extremes lie systems that, somewhat paradoxically, follow rules containing the seeds of their own disruption. The potential for dramatic change lies within them but remains dormant. All it takes is a little nudge, typically an imperceptible stress, to push them over the brink. We speak of a tipping point, a phase transition, the straw that broke the camel's back. Transformations like this can be surprising and logical at the same time.

Here the relevant mathematics is known as bifurcation theory. As the term suggests, a "bifurcation" is a splitting, a fork of possibilities. More precisely, a bifurcation is a qualitative change in a system's dynamics as some parameter is varied continuously. The contrast here is meant to be striking: the conditions (the parameters) change continuously, yet the resulting behavior changes discontinuously. You gradually turn up the heat and nothing happens until you reach the bifurcation point. Then the pot begins to boil.

∫ ∫ ∫

Joff's letters kept coming, once a month, sometimes twice. It got to the point where I stopped opening them. They'd lie on my desk, off to the side, a growing pile of envelopes addressed in that familiar cursive. Until one day a letter arrived whose cursive looked ragged.

Dear Steve, Sat. January 17, 2004

Eek! I had a mild stroke Thurs. noon and lost all sensation in my right (writing) hand. Several hours later I managed to open and close my fingers and get some strength back into my grip, but, alas, no dexterity! X@%X! A one-handed piano player isn't in demand, so I'll miss my gig with our jazz quartet tomorrow.

Just after I had completed the "rehab" of my new, titanium knee, I'm challenged to get control of my individual fingers. It is coming along; my middle and 4th fingers are holding back my progress. *Trying* to play the piano is helping, but frustrating.

My winter kayaking should work O.K.—as soon as this cold spell passes. Right now the saltwater cove I launch in is frozen too thick to get through.

Last week, while I was paddling, the cold icy water surely was giving my boat the "slows." I thought of Reynolds numbers and decided to revisit my physics book, then try to show that the complex fraction of units collapsed to a dimension-free number.

This led me to a pleasant retreat through viscosity, stress and strain in solids, then back to liquids, and Poiseuille. . . .

More physics encounters; this time with friction. I had discovered 8′ posts that I could use, in the golf course dump. I dragged the first two laboriously through the woods, which included an uphill stretch through a boulder "patch." That night I had an idea to use my mountain bike, which I use to tow my kayak, to drag the posts the long way on the edge of the course, then through a level section of woods to our mini-barn. It took a couple of rides through 3″ of snow to discover that the least friction came when most of the weight of the post was on my rear wheel and the tail of the post dragged in the snow.

It's evening now. At midday I rode my bike to my kayak launch site to affirm that this cold spell had frozen my access to the open waters of the Sound. Attempting to ride a farm road to the beach, I crashed when I skidded on an icy ridge covered by snow. Luckily I went down on my (good) left side. I'm used to banging shoulders; Sue and my sons are wishing I was more the type to relax by the fire with a good book. This seems like a reasonable hope.

Sunday Noon

Good news today: my self-invented rehab exercises are yielding encouraging results with my piano playing. Writing this finds me trying to get a comfortable grip on the pen.

Actually I really am reading a fascinating well-researched book about Magellan's circumnavigation (without charts or longitude) of thc earth. My grade school teacher gave me no hints as to the hardships involved. At least on Shackleton's trials in the Antarctic, he knew where he was. The Magellan book is *Over the Edge of the World*. I'm glad not to have been a sailor in Magellan's fleet.

Hope all is well with you, Carole, Leah, and Joanna. Sue and I are looking forward to our usual escape from the Conn. winter, to her family home on Hanalei Bay, Kauai.

Best regards,
Joff

I didn't write back to Joff about his stroke. Or call him either.

Maybe it was a kind of exhaustion. My dad had died a few months earlier, in October 2003. Watching him deteriorate was awful. So tiny in that hospital bed, his lips dry, always wanting more ice chips, his eyes unfocused, the humilation of the feeding tube. He'd never been a man of many words. My mother, whom he'd loved with every molecule in his body, had generated enough conversation for both of them.

I don't know if that had anything to do with why I didn't write back to Joff. But eventually I did contact him. It took something else to precipitate it.

In April 2004 my brother Ian, 57 years old, died suddenly one night. Same as my mom: stomach pain, emergency room, gone.

As soon as he heard about it, Joff sent me a note of condolence. Which underscored that I'd never done the same for him. It was time for me to change.

Dear Steve, Monday, April 19, 2004

I got a letter from Loomis with the news that your brother, Ian, had died. Sue and I are very sorry to hear this. My memory may be off, but I wondered if Ian had played JV football when Don Polkinghorne and I coached the team. The deep recesses of my mind have coach of the backfield, Don Polkinghorne, calling "Ian," but this singular recall may not have been accurate. In any case, know that we Joffs have a lot of sympathy for your loss, and a loss to the Loomis family of graduates.

Retirement from Loomis hasn't been a complete separation. In my dreams (NIGHTMARES?) I struggle to find my classroom as the time for the start bell to ring approaches! X@%X!

Please give our best regards to Carole, Leah, and Joanna.
Joff

I called him the minute I finished reading his note.

"Hello?"

"Hi Joff, this is Steve Strogatz."

"Oh, Steve." His voice descended mournfully when he said my name.

After thanking him for his condolences about Ian, I steered the conversation to his stroke. It was light, he said. He'd been shoveling snow. After finishing, he'd come in and sat down. Ten minutes later he felt something pop, "like a hiccup." His right hand fell asleep and did not wake up. Sue brought him to the doctor, who informed Joff he had an enlarged heart ("What do you expect after all those years of sports?" he chuckled in reply) and high blood pressure, and that his heart was a little leaky ("I never had a boat that didn't leak").

Then he changed the subject back to our usual math problems. Suppose you put 100 strands of spaghetti into a pot. If you tie the strands together at random until no loose ends remain, what's the total number of loops you expect to form? Laughing again, he said this question, posed by another of his former students, had been frustrating the hell out of him.

At the end of the conversation, something made me ask if I could visit him at his home for an afternoon, maybe sometime in the summer. We always visit Carole's mother in the city in August, I said, and it would be an easy drive up and back. Joff was tickled by the idea.

∫ ∫ ∫

On the morning of August 17, I drove north out of New York City, heading up Route 95 toward Old Lyme with a pocket tape recorder and a sense of trepidation. A few days earlier, I'd asked Joff if we could talk about some personal things we'd never discussed. He agreed, a bit reluctantly it seemed, but yes, he said, okay.

There were so many things I wanted to ask about but had always avoided. Jeff's cancer. Marshall's death.

I made a conscious effort to prepare my senses. Whatever was going to happen, I wanted to take it all in. I suppose it's pretty obvious by now that I'd lived much of my life in my head. Most of it, really. But on this day, I told myself, I'm going to open my eyes and see Joff for the first time, and listen to him.

$\int \int \int$

After driving for a little over two hours, I got off the highway at Exit 70 and navigated the back roads to Joff's street. It was a dead end, with five houses on the left, all perched on a small hill. On the right was a marsh, with dozens of birdhouses standing on wooden poles poking out of the wetlands. Each one was numbered, with addresses like $\sqrt{e}$, π, and other silly constants.

I parked at street level. As I walked up the driveway, I heard the soft sound of a piano from inside the house. The screen door was ajar. "Hello?" The piano stopped instantly. Sue and Joff greeted me, both looking tan and youthful at 75. I kissed her on the cheek. Joff and I hugged, gripping each other on the backs of our shoulders.

Sue took me on a tour of the house. I noticed a wall filled with photos and felt my breath catch in my throat. Even though I didn't look at them, I felt myself starting to choke up—I sensed they were pictures of their three sons and their families. Sue took me upstairs to her studio to look at her paintings, mostly still lifes. On the way back downstairs, I went past more walls full of family pictures, and didn't look.

Joff took me down to his basement workshop. Canoes, kayaks, and windsurfing boards hung on the back wall above cinder blocks on the floor. The smell—oil? mustiness?—reminded me of workshops from my childhood. He said he only has simple tools: a band saw, drill press, another kind of saw, but not a rotary saw. "Why do I need a rotary saw?" he asked with a smile. He said he was proud of the windows that overlook the marsh and let in a lot of light. He told me stories about the $\sqrt{e}$ numbering, and about his excursions on kayak to repair the nest boxes when the marsh is flooded high enough in the spring.

Our next destination was the porch deck. We sat under a big umbrella and had lunch: cold cuts and various breads followed by cherry pie with vanilla ice cream.

When it was finally time for our conversation, I took out the tape recorder. Joff began leafing slowly through his personal journal, which included math problems inspired by his outdoor activities. We went through it page by page, looking at his drawings of birds seen at this home or in Hawaii. Then lots of stories about his friend Hank, the bird guru.

I wonder, will this go on and on? I fight with myself, about my own impatience. This matters to him, so pay attention. I try to.

Eventually we got talking about Jeff, who had grown up to become an engineer. He had had testicular cancer in his early twenties, battled through surgeries and chemo, and made a great recovery. Afterward he appeared on television, snowboarding down a mountain for a Channel 3 newscast, saying he beat testicular cancer. Hearing about his miraculous recovery made me feel let off the hook for my longstanding callousness.

After a lull, I asked about Marshall. I'd been waiting to find the right time. My words spilled out.

"So I don't think we ever talked about Marshall, but I wanted . . . I . . . I didn't really know him . . . but he . . . I know that he died very young, and I . . . what, what happened to him?"

"Well, we, you know, that's something we don't really. . . ."

"You don't want to talk about that?"

"Ahh, well, it was a sad. . . . He, he had. . . ."

"I remember him as a star. . . ."

Our words were colliding now. I'd gone too far. Feeling frantic, I looked for a way out. Then, unexpectedly, Joff began.

"He had a wonderful 27 years. He went to Mannes. He started at Amherst but music was gonna be his thing. He was interested in concert piano. And so he went to Mannes. . . ."

". . . Which is a music school?"

"Yeah. It's near Juilliard. It's in that complex of music schools. It's not Juilliard but it's what he could get into. And, oh, he loved it. He had a little apartment down there, and we used to go down and visit him. And, uh, then . . . oh, he got this illness which was the thing that ultimately did him in."

We sat quietly.

"And, it was a sad . . . because even, even in his waning moments, he'd stay up all night long, playing the piano, and just . . . the house would be filled with beautiful music. And he had made plans to get a job at the New England Conservatory and things like that, but the fates were wrong for him. But at least, I don't know anybody that had . . . more . . . interesting things in that 27 years. I mean, he starred in every musical production the school theater ever had at Loomis. He sang in Jim Rugen's madrigals. He composed—I was going through the files the other day and he'd written this kyrie, a religious piece that had been recorded."

"Did he have a religious feeling?" I asked.

"Mmm. . . ."

"Not really?"

"Mmm, yeah. I think he did. I think he felt close to having to come to terms with somebody out there."

Joff fell silent. We sat quietly and looked out over the salt marsh.

"So, I guess, that was a good thing, that, that, I think he went peacefully. . . . Oh yeah, we miss him. He traveled all over. A friend of his said, 'Hey, I just won tickets to Israel. You want to come? I've got two of them.' So he was off there. He just . . . he had a great time."

Neither of us spoke.

"He was in a boy choir, starting when his voice was little. And at the church in Hartford, he was the head boy in the choir, and their choir went over to Westminster Abbey and sang one summer. I'd always admire his discipline. You know, a lot of great things happened. He was quite a scholar. When Sue was studying for her master's degree at Wesleyan, she'd call him up and ask him questions. He was in the business of researching some of these lesser known composers. The guy was just a storehouse of information. And I always envied him the trick that he'd be at home and we'd sit around the piano and he'd say, 'Okay, what'll we do?' And I'd get out the Cole Porter songbook and just turn to a page, something that he'd never seen. He could sight-read it, play it, and sing it all at one time with us. And I thought, 'God, this guy's got a multichannel mind that I wish I had!' "

I asked about one of Marshall's friends I'd known, and before long the conversation returned to easier topics and calculus problems.

After a while Joff asked if I'd like to go to the beach for a swim, or maybe just stroll by the water. We found a nice patch of sand and discussed a calculus problem that one of his former students had sent him. It was a problem about waves and required a Fourier integral—a generalization of Fourier series with uncountably many sine waves—for its solution. This was a different kind of infinity than Joff had ever encountered before, a higher order of infinity. While I explained it to him, the sun began to set. We sat together on the beach and solved the problem, surrounded by the waves in Long Island Sound.

∫ ∫ ∫

We made our way back to his house as the sky darkened. He and Sue fed me a quick dinner—leftovers from lunch—before my ride back to the city.

I kissed Sue and hugged Joff goodbye. As I was stepping out the door, he handed me an envelope with my address on it.

I climbed into the car and opened the letter. At the end of it, he'd written

---

It just dawned on me that I could save postage and hand-deliver this letter when you arrive. I confess being nervous about seeing you. Our friendship/correspondence has meant so much over the years. Just the thought of you braving the drive up 95 from New York City for a visit is overwhelming.

Best wishes for a safe passage!
Joff

He was nervous about seeing me. Unbelievable. I was nervous about seeing *him*. What was *he* nervous about?

Hero's Formula
(2005-Present)

Calculus began with Zeno's paradoxes about time, motion, and change. The most famous one is about Achilles and the tortoise. Consider a race between the great warrior and a humble tortoise in which the creature is given a head start. By the time Achilles reaches the tortoise's starting place, the tortoise has crawled a bit farther ahead. By the time Achilles gets there, the tortoise has moved a little more and hence is still in the lead. Thus, a swift runner can never overtake a slower one. Since common sense says otherwise, Zeno concludes that common sense is wrong. Change is an illusion. You should trust your mind and nothing else.

Most mathematicians will tell you that Zeno was mixed up about infinite series. Now that we understand calculus, Zeno can be dismissed. Still, I feel some sympathy for what he was fretting about. Looking through my correspondence with Joff makes me feel acutely aware of the past gaining on the present, the years rushing up from behind. Here in the slow-moving present, Joff and I are like tortoises chased by time.

He knows that I'm writing about us. Whenever we speak on the phone and the subject of the book comes up, he says he just hopes he'll live to see it.

∫ ∫ ∫

Back in 2007, we hadn't been in touch for about two years. On February 15, 2007, I wrote to apologize for how long it had been. A few days later Sue called. "Don had a second stroke," she said. "It blinded him in both eyes. But only to the right; his vision to the left is fine. He can't drive anymore, but he still bicycles his kayak out to the river, and even though his quick mind isn't quite there, he still remembers all the old stuff. Shall I put him on?"

Sure, I said, I'd love to talk to him.

I was scared.

Joff picked up, his voice warm as ever. "Don't worry," he reassured me. "Even though I can only see about 7 degrees to the right, I just think, here comes that right hand gate and turn my head. My neck is still flexible from all those days of negotiating reverse gates in whitewater slalom races."

A few months later I wrote him a short note, asking for permission to reproduce all his letters. His reply, written in an unsteady hand, contained a few cross-outs that clearly embarrassed him.

Dear Steve, Monday Nov. 5, '07

My short-term memory got altered in my last stroke, but I've been able to find my way home from my bike rides and kayaking in local waters.

I'm so pleased that you are writing a story of our correspondences over the years. I *give you complete authorization to write about the correspondences we had over the years*. There were so many times that you responded to my questions about math notions that puzzled me. My students came to class and their first questions were "Any letters from Steve today?"

I can't believe that our letters were not singularities in our teacher-turned-student relationship.

As you can see, writing this thank-you letter has required some alterations. My hope is to say a heartfelt thank you for these years.

(I'll ask Sue to proofread this letter!)
Sincerely, Don

While writing this book, I've been thinking a lot about what I learned from Joff. For years I would have said, not much, meaning not much math. That was true even in high school. But I'm starting to realize what it was that he gave me.

He let me teach him.

Before I had any students, he was my student.

Somehow he knew that's what I needed most. And he let me, and encouraged me, and helped me. Like all great teachers do.

But now I also see that I did learn something from him, something profoundly mathematical, about how to live. From his hobbies to the way he faces the ups and downs in his life, Joff is brave about change. He rolls with it and tries to make peace with it. And where he can, he even plays with it. Jazz piano, windsurfing, whitewater kayaking—all of these balance the inevitable against the unforeseeable, the two sides of change in this world. The orderly and the chaotic. The changes that calculus can tame, and the ones it cannot. He confronts them all, and not, like Zeno, with his mind alone but also with his heart.

∫ ∫ ∫

Joff wrote his last two mathematical letters to me in 2005, just before his second stroke. The problem that intrigued him was a classical question in geometry: how to find the area of a triangle given the lengths of its three sides. For a right triangle this is easy, but for one with arbitrary angles the answer is a formula that nobody teaches anymore, a result known as Hero's formula, after the Greek mathematician who discovered it.

These old letters make me smile. Not so much because of the math they contain but because of the Joff I see once again on the pages—his cartoon drawings, his self-flagellation and outbursts of happy exasperation, and his pure pleasure in the logic of the argument.

Dear Steve, Saturday 4 June '05

I thought you'd get a smile from hearing about a recent happening. A former Loomis Chaffee graduate sent me a letter which included her notes on a talk she gave demonstrating Hero's proof of his area of a triangle in terms of its sides. It wasn't surprising that the proof had a lot of geometry. (What the heck is a cyclic quadrilateral?, I wondered.)

I read part way thru her notes, but decided it would be fun to attempt my own proof using a modern approach with trig.

After pursuing a few paths that led me nowhere, or elaborately writing a lot of lines that got me back to my starting point, I came up with what I hope will survive scrutiny. Now if you have an idle moment, you might enjoy my mathematical meanderings in trying to construct a proof.

At Loomis, I believe the math teachers accepted $\sqrt{s(s-a)(s-b)(s-c)}$ without proof. Some of my retired friends do crossword puzzles to keep their brains exercised, but that activity never captured my enthusiasm. My *old* geometry teacher at Wilbraham confided to me that there was nothing more satisfying to him than spending an evening on proving a geometric original. (This seemed strange to this 15-year-old who was heavily into sports.)

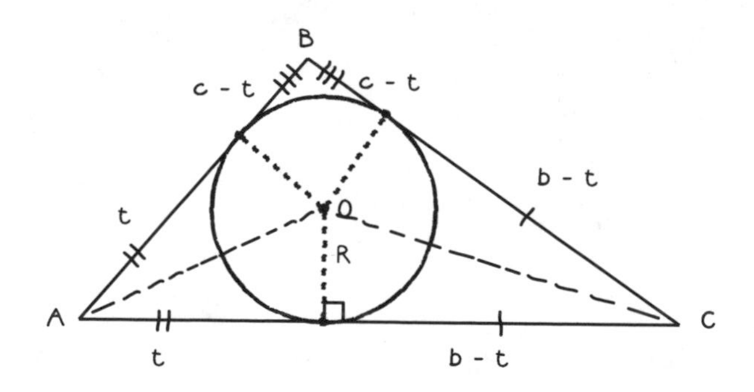

Task: Show area $\triangle ABC = \sqrt{s(s-a)(s-b)(s-c)}$, where the semiperimeter $s = \frac{1}{2}(a+b+c)$. Here O is the in-center of $\triangle$, so area $= Rs$ is easily established, so I went to work on tangent lengths:

$$2s = 2t + 2(c - t) + 2(b - t)$$

$$\begin{cases} s = b + c - t \\ 2s = b + c + a \end{cases}$$

$$s = a + t; \qquad t = s - a;$$

this extends to giving the other tangents as cyclic permutations of a, b, c.

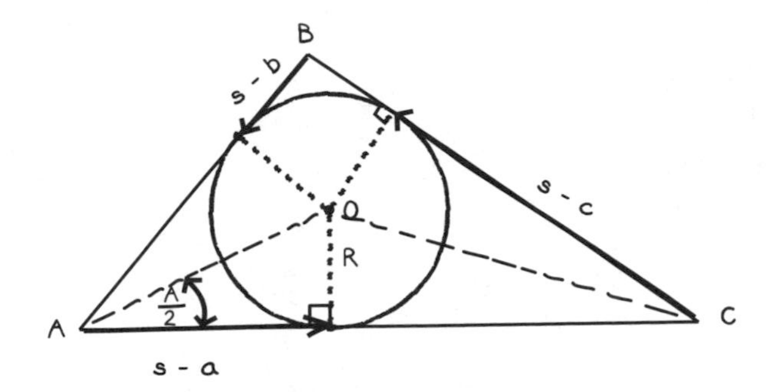

$$\left.\begin{aligned} \tan\frac{A}{2} &= \frac{R}{s-a} \\ \tan\frac{B}{2} &= \frac{R}{s-b} \end{aligned}\right\} \tan\frac{A}{2}\tan\frac{B}{2} = \frac{R^2}{(s-a)(s-b)}.$$

Also,

$$\tan\frac{A}{2}\tan\frac{C}{2} = \frac{R^2}{(s-a)(s-c)}$$

and

$$\tan\frac{B}{2}\tan\frac{C}{2} = \frac{R^2}{(s-b)(s-c)}.$$

Adding,

$$\begin{aligned}&\tan\frac{A}{2}\tan\frac{B}{2}+\tan\frac{A}{2}\tan\frac{C}{2}+\tan\frac{B}{2}\tan\frac{C}{2} \qquad \text{(Equation (I))}\\ &=R^2\left[\frac{1}{(s-a)(s-b)}+\frac{1}{(s-a)(s-c)}+\frac{1}{(s-b)(s-c)}\right]\\ &=R^2\left[\frac{s-c+s-b+s-a}{(s-a)(s-b)(s-c)}\right]\\ &=\frac{R^2 s}{(s-a)(s-b)(s-c)}.\end{aligned}$$

Interlude:

$$\frac{A}{2}+\frac{B}{2}=90-\frac{C}{2}$$

$$\tan\left(\frac{A}{2}+\frac{B}{2}\right)=\frac{1}{\tan\frac{C}{2}}$$

$$\frac{\tan\frac{A}{2}+\tan\frac{B}{2}}{1-\tan\frac{A}{2}\tan\frac{B}{2}}=\frac{1}{\tan\frac{C}{2}}$$

$$\tan\frac{A}{2}\tan\frac{C}{2}+\tan\frac{B}{2}\tan\frac{C}{2}=1-\tan\frac{A}{2}\tan\frac{B}{2}$$

$$\tan\frac{A}{2}\tan\frac{B}{2}+\tan\frac{A}{2}\tan\frac{C}{2}+\tan\frac{B}{2}\tan\frac{C}{2}=1.$$

Neat! (Maybe I should have known this triangle property, but it was nice discovering it.) Returning to Eq. (I),

$$1=\frac{R^2 s}{(s-a)(s-b)(s-c)}$$

$$R^2=\frac{(s-a)(s-b)(s-c)}{s}, \qquad R=\sqrt{\frac{(s-a)(s-b)(s-c)}{s}}$$

$$\text{Area } \Delta = Rs = s\sqrt{\frac{(s-a)(s-b)(s-c)}{s}}$$

$$=\sqrt{s(s-a)(s-b)(s-c)}$$

Phew! X%@!

Retired life seems busier than ever. A list of projects to be done during a day still has items not crossed off at the day's end, and that's good. Next Saturday's plan of the day finds me giving a short speech at the Wilbraham chapel about my former physics and chemistry teacher, Phil Shaw, who also was my track coach. His inspiration made me choose the prep school career.

That evening my LAIDBACK JAZZ TRIO plays at the LC reunion cocktail/reception party from 7 to 8. The next morn our trio plays a brunch gig in Old Saybrook from 11 to 2. [Hurry sundown?]

I got an early start sailing this year, despite the cool weather and water. Kayaking and biking round out my escapes to the serenity (soon to change) of Long Island Sound. At least the tidal creeks won't be invaded by the power boaters.

Oh, well, lifestyles have become very complex these days. I'm lucky to be able to avoid some of the confusion.

Back to the cartoon on page one . . . I certainly can appreciate the genius of Hero . . . and that remarkable group of Ionians of the Alexandrian period, Democritus, Aristarchus, Eratosthenes. . . .

Best regards, to you, Carole, Leah, and Joanna,
Joff

Hi Joff, 6/19/05

Your letter with the terrific proof of Hero's formula found its way to me here at the Niels Bohr Institute, where Carole,

Leah, Joanna, and I have been spending a *wonderful* sabbatical since January.

So much to love about life in Copenhagen—it's kid-friendly, with play spaces in every bank or business. Everyone speaks perfect English and seems to like Americans. They have unbelievable public transport and facilities; for instance, our local swimming hall is an Olympic-size pool complete with marble statues all around it.

Anyway, we'll be very sad to leave—and our departure is in just two days. !?@X! as you would write.

I've been able to get some fun research done, since the phone never rings here. The problem that consumed me for 6 months was the wobbling of London's Millennium Bridge (which I wrote about in *Sync*)—how it was caused by the pedestrians on it when they inadvertently synchronized their paces to the bridge's slight swaying, thereby amplifying the wobbles.

The atmosphere of this historic place—it's actually Bohr's house and a building next to it—is fantastic. I was thrilled to be asked to lecture in Auditorium A which has been perfectly preserved as it was in the 1920s, when Bohr, Heisenberg, Pauli, Dirac, and other luminaries sat in that same little room and argued about the newly invented "quantum mechanics."

OK, well, have to keep it short—we're packing furiously and readying ourselves for reinjection into the hurly-burly of American life. Hope to write more soon, from the usual Kimball Hall address. But I thought you might get a kick out of the unusual envelope and return address, so I couldn't resist firing off this quickie!

Cheers,
Steve

Dear Steve, Tuesday 7/26/05

Wow! An "A Prioritaire" letter from NIELS BOHR INSTITUTET!?!! That is indeed a coveted postmark. Did you learn Danish on your sabbatical?

This correspondence is a blushing confession about my mental activities at age 76. My friend, Nina, who sent me her notes from a talk about the Hero formula, also included an epilogue entitled, INTERESTING THINGS. One was a remark that the proof she'd presented may have come from Archimedes. Another commented that one could "indirectly prove the Pythagorean theorem from the Hero area formula."

The latter claim sent me back to $A = \sqrt{s(s-a)(s-b)(s-c)}$ to show the Pythagorean theorem was embedded in the formula. I started with a right triangle:

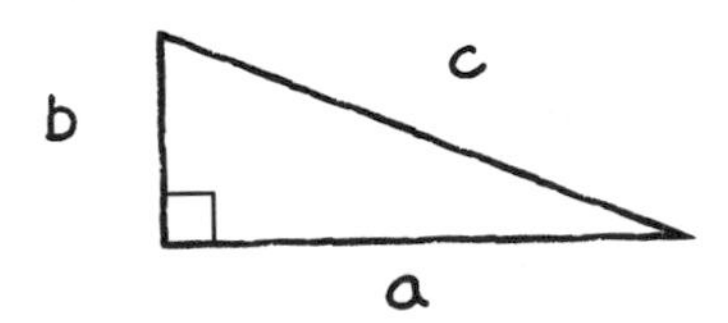

then assumed its area was $\frac{1}{2}ab$, and, after laboring thru the algebra of sides and semiperimeters, got $a^2 + b^2 = c^2$ to fall out of the formula. Not exactly elegant stuff. As for a *proof* of the Pythagorean theorem, there's probably circularity in this work, but I haven't checked out the Hero's (Archimedean?) proof to see if it used the Pythagorean theorem. No, I rushed off to more embarrassing pursuits.

One day I recalled another note in Nina's INTERESTING THINGS epilogue: "It is proven that quadrilateral ABCD has area

$$A = \sqrt{(s-a)(s-b)(s-c)(s-d)} \text{ ''}$$

and "make this quad into a triangle by letting $d = 0$, then

$$A = \sqrt{s(s-a)(s-b)(s-c)}. \text{ ''}$$

My letter deserves a discontinuity here. Please be patient (a lot of physicists must have had to be patient with Max Planck's taking them to new energy levels).

A single gal, a neighbor down the street, has asked me to build a boat port for the 15′ rowboat she built last winter. Since I had a lot of scrap lumber, I made some sketches of the structure needed to support a tarpaulin she bought.

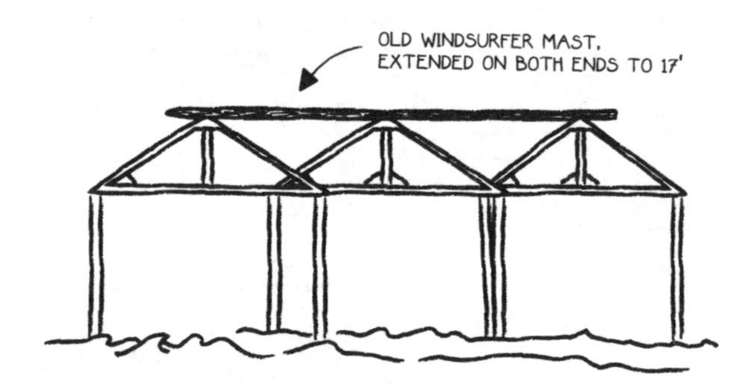

I *knew* a quadrilateral was not a rigid figure, so you can see I was "prefabbing" a lot of triangles into the structure.

OK, back to my confession. (Assume my brain activity has descended several energy levels.) I got out pencil and paper and attempted to prove area of quad was $\sqrt{(s-a)(s-b)(s-c)(s-d)}$. Checking to see all checked out for a square, and assuming the unspecified s was again a semiperimeter $s = \frac{1}{2}(a+b+c+d)$, I filled a lot of pages working with adding to triangles sharing a common side, their semiperimeters differing.

After a lot of algebra which came up short, I wondered if testing a rectangle would be prudent. Ahhh, the formula survived, but at that instant a light bulb flashed (*low* wattage, surely no more than 40 watts). My rectangle could easily deform to a parallelogram with the same semiperimeter. I sketched a cartoon to chide me for my attempts to prove the area of a quadrilateral could be reckoned from its sides:

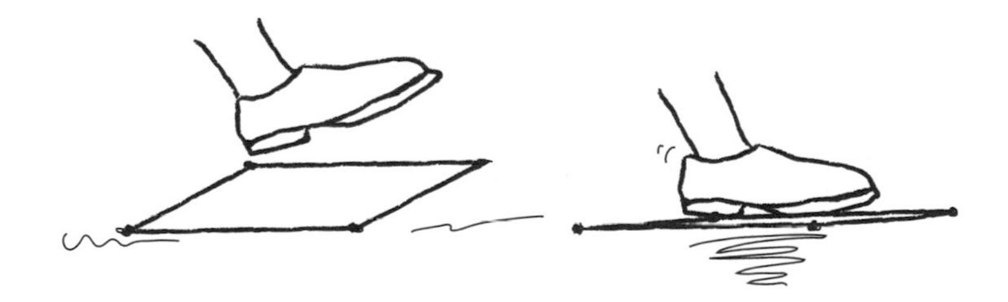

Arghhhhh! %@X!

Maybe some teachers' publisher might have the delight to introduce a magazine of TRUE CONFESSIONS (of teachers).

I might be a source of anecdotes:

One spring I came to a pause in a multivariable calculus proof on the board. Clearly struggling for the next step, I saw my student, Adam Doctoroff, raise his hand . . . and he gave me the next step. I praised him for the help and asked, "Where did you come up with that?" His response, "Oh, you showed me that last fall."

Hey, Steve, I was fortunate in my career at Loomis to have students with superb minds . . . including a young man who, at LC, was fascinated with Huygens, and, at a student comedy show in the Chaffee gym, donned a white lab coat and did a chalkboard proof that $2 = 1$.

Mm, time to get back to building my neighbor's boat port.

I love you, Steve.
Joff

Acknowledgments

A friend of mine once ribbed me about how I approach everything—"with all cards on the table, face up, all the time." Even so, it wasn't easy to be as honest as this book required. To whatever extent I've gotten it right, I owe it to the encouragement and support of my colleagues, friends, and family.

Thanks to Sam Arbesman, Karen Dashiff Gilovich, Tom Gilovich, Anna Pierrehumbert, and my mother-in-law Shirley Schiffman for reading the manuscript and offering a wide range of helpful comments.

June Meyermann, technical typist extraordinaire, was the book's first reader. Her cheerful impatience to receive the next batch of pages was what any author longs for.

Louise Moran, director of communications at Loomis Chaffee, rummaged through the archives and kindly supplied me with excellent photographs of Joff.

Margy Nelson prepared the artwork with her signature combination of precision and whimsy.

Will Schwalbe, who edited my last book, urged me to follow my dreams for this one. Thanks for your advice at the earliest stages and for steering me in the right direction.

My literary agent, Katinka Matson, also believed in this book from the start and helped it find a happy home. Her instincts are always terrific, and matched only by her candor and sense of humor.

Alan Alda, what a good friend you've been. Thanks for the many brainstorming sessions and for all your invaluable advice about dramatic action (a concept we never covered in math class). Whenever I was stuck, you were the reader I would imagine writing for.

Thank you to Vickie Kearn, my editor at Princeton University Press, for your insights (always gently delivered), your sunny enthusiasm, and your unflagging support in all phases of this project. It was wonderful working with you.

To my wife Carole, my best friend and love of my life, you've heard it a hundred times already, but for posterity: thank you so much for your sympathetic and shockingly incisive advice about how to improve the earliest drafts of the manuscript. And thanks for tolerating my obsession with this project and for understanding why it mattered so much to me, even before I knew why myself.

Above all, thank you to Don Joffray. I can never repay you for the warmth of your friendship, for everything you've taught me, for all the mathematical fun we've had together, and for so generously allowing me to share our correspondence. But I do have a great calculus problem I can't wait to tell you about. I'll call you soon.

Steven Strogatz
Ithaca, New York
July 14, 2008

Further Reading

Here are some suggestions for further reading, organized chapter by chapter, for anyone who'd like to learn more about the mathematical topics that Joff and I discussed in our letters or that arose elsewhere in this book.

I've tried to recommend books that are friendly and accessible. Most require a knowledge of first-year calculus but not much beyond that.

For convenience, books are cited briefly here by author and year of publication. More detailed information appears in the bibliography.

Continuity

Dunham (2004) traces the development of rigor in calculus, including the modern definitions of limits and continuity.

The Fibonacci numbers and their connections to nature, art, and architecture are engagingly discussed in Livio (2002) and Posamentier and Lehmann (2007).

Pursuit

Nahin (2007) is the book for one-stop shopping about chase problems, complete with their history, analysis, and applications.

Relativity

Isaacson (2007) is a perceptive and beautifully written biography of Einstein. It also does a fine job of explaining the theory of relativity and Einstein's other scientific contributions in laymen's terms.

Martin Gardner's formulation and calculus-free solution of the four-dog chase problem can be found in Gardner (1994); see Problem 16, "The Amorous Bugs." The interesting back story of this problem, including Gardner's role in popularizing it, is discussed in Chapter 3 of Nahin (2007).

Irrationality

Maor (2007) presents a delightful account of the Pythagorean theorem and its larger significance. The simple proof given there (in Appendix D) demonstrates that the square root of *any* prime number, not just 2, is irrational.

Shifts

The Fibonacci numbers satisfy a linear difference equation with constant coefficients. Goldberg (1986) and Elaydi (2005) present readable introductions to difference equations and their applications.

Proof on a Place Mat

For the amusing story of how Feynman learned to differentiate under the integral sign and how valuable that technique proved to be later in his career, see the chapter entitled "A Different Box of Tools" in Feynman (1985). Incidentally, if you haven't already read that book, it is wonderful—essential reading for any budding physicist.

Feynman says he learned advanced calculus by studying Woods (1926). This classic text contains much juicier examples than the ones we offer our students nowadays. Differentiation under the integral sign is considered on pp. 141–163.

Keener (1988) and Simmons (1991) offer nice introductions to the gamma function.

The Monk and the Mountain

The easiest place to find Martin Gardner's version of the monk and the mountain puzzle is in Gardner (1994), where it appears as Problem 50, "A Fixed-Point Theorem." Also, see his solution and comments on p. 74. Gardner originally wrote about this brainteaser in his Mathematical Games column in *Scientific American* in June and July of 1961. He says the puzzle was brought to his attention by the psychologist Ray Hyman, who in turn found it in a monograph entitled *On Problem Solving* by Karl Duncker, a Gestalt psychologist. The problem was given a further boost in popularity when it appeared in Koestler (1964). Since then it has been a staple in psychology courses and books on creativity. For instance, see the bestseller by Adams (2001).

Dimensional analysis is beautifully explained and illustrated with many real-world examples in McMahon and Bonner (1983). For the use of dimensional analysis in mathematical modeling, see Bender (1978) and Lin and Segel (1988).

Wallis's formula is discussed, both mathematically and historically, in Section B.12 of Simmons (2007).

Randomness

The Monty Hall problem was fresh at the time that Joff and I were discussing it, but lately it has come to seem a bit hackneyed, having been overexposed in many popular books on probability. Instead of any of those, I'd recommend that you consult the web site for "Ask Marilyn" (http://www.marilynvossavant.com/articles/gameshow.html), where you can find the hilariously irate letters excoriating her, along with her calm replies. Or take a look at the moving and wonderfully quirky novel by Haddon (2003). The Monty Hall problem is discussed there on pp. 62–65.

Infinity and Limits

The limit that puzzled Joff appears on p. 384 of Boyer and Merzbach (1991).

Laplace's method for the asymptotic evaluation of integrals is explained in Section 6.4 of Bender and Orszag (1978) and in Section 10.3 of Keener (1988). Honestly, understanding these will require some serious effort unless you've taken a few college courses beyond calculus.

Chaos

For a superb popular account of chaos theory, see Gleick (1987). For an introduction to the mathematics of chaos and its many scientific applications, see Strogatz (1994).

The infinite product discussed in this chapter is known as Vieta's product and is proven in Section B.9 of Simmons (2007).

Celebration

The Winter 2000 edition of *Loomis Chaffee Magazine* contains two long and affectionate articles, tributes to Don and Sue Joffray upon the occasion of their retirement. The text of my speech at St. John's Cathedral appears on pp. 11–13 of that issue.

The Path of Quickest Descent

For a gentle introduction to the cycloid and its tautochrone and brachistochone properties, along with interesting historical comments, see Sections B.21 and B.22 of Simmons (2007).

Bifurcation

The basic ideas of bifurcation theory are discussed by Gleick (1987), and in more mathematical form by Strogatz (1994).

The book about Magellan that Joff mentioned in his letter of January 17, 2004, was Bergeen (2003).

Hero's Formula

For Zeno's paradoxes, see Mazur (2008).

Hero's formula (also known as Heron's formula) is proven and discussed by Simmons (2007) in Section A.7 and on pp. 222–225.

Simmons also points out that the natural generalization of Hero's formula to quadrilaterals—which Joff chided himself for believing could possibly exist—actually does exist! The key is to require that the quadrilateral be inscribed in a circle. With that constraint, the formula Joff struggled to prove really is true. It was discovered in the 7th century by the Indian mathematician Brahmagupta and is now known as Brahmagupta's formula. See Simmons (2007), pp. 222–225.

Bibliography

J. L. Adams, *Conceptual Blockbusting: A Guide to Better Ideas* (Basic, New York, 2001), pp. 4–5.

C. M. Bender and S. A. Orszag, *Advanced Mathematical Methods for Scientists and Engineers* (McGraw-Hill, New York, 1978).

E. A. Bender, *An Introduction to Mathematical Modeling* (Dover, Mineola, New York, 2000).

L. Bergreen, *Over the Edge of the World: Magellan's Terrifying Circumnavigation of the Globe* (William Morrow, New York, 2003).

C. B. Boyer and U. C. Merzbach, *A History of Mathematics,* 2nd Edition (Wiley, New York, 1991).

W. Dunham, *The Calculus Gallery: Masterpieces from Newton to Lebesgue* (Princeton University Press, Princeton, New Jersey, 2004).

S. Elaydi, *An Introduction to Difference Equations,* 3rd Edition (Springer, New York, 2005).

R. P. Feynman, *"Surely You're Joking, Mr. Feynman!": Adventures of a Curious Character* (W. W. Norton, New York, 1985).

M. Gardner, *My Best Mathematical and Logic Puzzles* (Dover, Mineola, New York, 1994).

J. Gleick, *Chaos: Making a New Science* (Viking, New York, 1987).

S. Goldberg, *Introduction to Difference Equations* (Dover, Mineola, New York, 1986).

M. Haddon, *The Curious Incident of the Dog in the Night-Time* (Doubleday, New York, 2003).

W. Isaacson, *Einstein: His Life and Universe* (Simon and Schuster, New York, 2007).

J. P. Keener, *Principles of Applied Mathematics: Transformation and Approximation* (Addison-Wesley, Redwood City, CA, 1988).

A. Koestler, *The Act of Creation* (MacMillan, New York, 1964), pp. 183–184.

C. C. Lin and L. A. Segel, *Mathematics Applied to Deterministic Problems in the Natural Sciences* (Society for Industrial and Applied Mathematics, Philadelphia, 1988).

M. Livio, *The Golden Ratio: The Story of Phi, the World's Most Astonishing Number* (Broadway, New York, 2002).

E. Maor, *The Pythagorean Theorem: A 4,000-Year History* (Princeton University Press, Princeton, New Jersey, 2007).

J. Mazur, *Zeno's Paradox: Unraveling the Ancient Mystery Behind the Science of Space and Time* (Plume, New York, 2008).

T. A. McMahon and J. T. Bonner, *On Size and Life* (Scientific American Library, New York, 1983).

P. J. Nahin, *Chases and Escapes: The Mathematics of Pursuit and Evasion* (Princeton University Press, Princeton, New Jersey, 2007).

A. S. Posamentier and I. Lehmann, *The Fabulous Fibonacci Numbers* (Prometheus, New York, 2007).

G. F. Simmons, *Differential Equations with Applications and Historical Notes,* 2nd Edition (McGraw-Hill, New York, 1991).

G. F. Simmons, *Calculus Gems: Brief Lives and Memorable Mathematics* (Mathematical Association of America, Washington DC, 2007).

S. H. Strogatz, *Nonlinear Dynamics and Chaos: With Applications to Physics, Biology, Chemistry, and Engineering* (Perseus, Cambridge, Massachusetts, 1994).

F. S. Woods, *Advanced Calculus: A Course Arranged with Special Reference to the Needs of Students of Applied Mathematics* (Ginn, Boston, 1926).

Photography Credits

Grateful acknowledgment is given for permission to reprint the following photographs.

On page 4: Loomis Chaffee Archives, 1974.
On page 83: Loomis Chaffee Archives, 1990.
On page 116: Leo Sorel Photography, 1999.
On page 153: Michael Chen, Loomis Chaffee Archives, 1992.

Index of Math Problems

For math students, teachers, and enthusiasts, here's an index of the problems discussed in this book. They're organized by subject, starting with more elementary topics and progressing to more advanced material.

A problem followed by the symbol "(S)" means the solution is given in the pages listed. Otherwise, the problem is posed but not solved here—you might like to investigate such questions on your own and, if you're a teacher, assign them to your students.

Geometry

Trigonometry

Probability and discrete math

Calculus

Differential Equations

Fourier series

Complex variables

Asymptotic methods

Calculus of variations